DISCOVERY WORLD

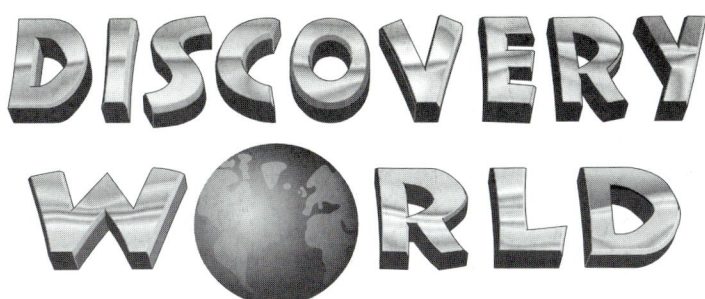

Skills Development and Assessment Guide

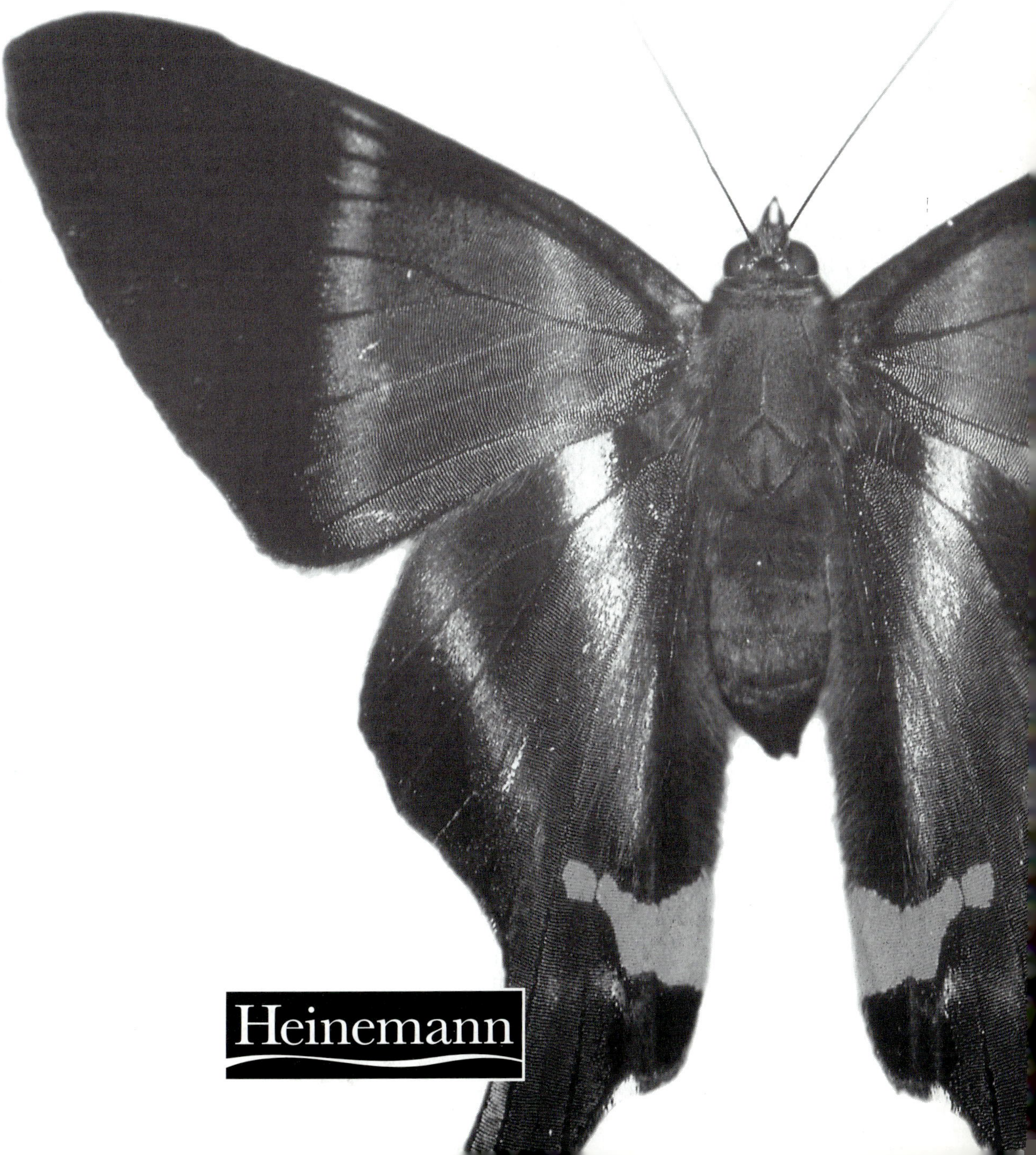

Heinemann

About this Guide

 This *Skills Development and Assessment Guide* has been designed to help the busy teacher.

It has been designed to help *you*.

 The main supporting teaching notes for **Discovery World** are actually found in *Literacy Lesson Books 1 and 2*.

*To make the programme work for you as a skills system you will need these books in **your** classroom.*

 This *Skills Development and Assessment Guide* has been designed for you to dip into to get the information you need as easily as possible.

You can turn straight to the section that meets your needs using this *Quick* contents.

Quick contents

 We chose the chameleon as our mascot and colour grading symbol because of its ability to adapt itself to any surroundings.

We hope that **Discovery World** helps you prepare your children for the fast-changing modern information age.

Contents

Welcome to Discovery World!

Discovery World is an exciting new programme designed to help children acquire essential non-fiction skills from ages 4–11.

Discovery World provides children's books that are:

- of the highest-quality design, to excite today's video-age child,
- graded, so even the youngest of readers can read and understand the text,
- models of many different types of non-fiction texts (biography, dictionary, diary, etc.),
- based on all the most popular topics and themes you teach,
- based on the content of the National Curricula.

Discovery World is supported by a unique skills system that:

- is written by teachers to be as user-friendly as possible,
- is based on the idea of 'literacy lessons',
- is based on 'modelling' strategies (adapted from the use of big books),
- contains strategies for differentiation,
- contains strategies to develop children's non-fiction writing.

Why is non-fiction important?

Children's early reading experience has traditionally been in the area of fiction. Research has shown that over 80% of reading time at age 4–7, including reading instruction, has been spent on reading fiction. In the past it was assumed that children would only be able to read non-fiction texts once they were proficient readers of fiction.

However, children's reading experience *outside* school is rich with information print. Some of the first books ever given to children in the home (such as baby board books or picture alphabet books) are often simple catalogues of labelled items. These are information books. Even the youngest of children can often read and recognise packaging, adverts and shop signs. Most teachers recognise this environmental print as valuable reading experience, but may still neglect non-fiction books in the classroom as an early route to literacy.

Children may spend their early years immersed in poems, traditional tales, stories and reading scheme 'readers'. They start to learn about the conventions of character and plot. They learn that every story starts on page one, reads from top left down the page, and has a beginning, middle and end. By the time children are aged about 7 their reading diet suddenly changes. Almost overnight, it seems, they are expected to read and understand non-fiction books, textbooks, and specialist information books about science, geography or history.

It is now recognised, therefore, that non-fiction books should be read and discussed by children from the earliest stages of reading:

- Their highly visual presentation can make them *easier* for the very young reader to access than fiction (e.g. connecting a photo of a postman with the simple line *What does a postman wear?* in *Special Clothes*).

- They are ideal for *shared* or *guided* reading since they contain images from the child's real experience, ideal for discussion.

- They offer a better route to literacy for some readers (often boys) who prefer the real-world content.

- The range of topics available provides a greater choice for readers of all abilities and interests.

The benefits of teaching young children early non-fiction skills from school entry to age 7 are considerable. It is a foundation for the skills they will need throughout their school life, and for their adult lives (most of the reading we do as adults is information reading). It will unlock other subject areas like Science, History and Geography, which rely on the use of non-fiction books and the interpretation of charts and graphs, timelines, diagrams, maps etc. Literacy research has shown that children's *reading* experience directly influences their ability to *write*. Experience of different types of non-fiction in the early years has been shown to assist children's ability to write for different audiences in an appropriate style later on.

Overuse of stories as the major reading experience in the early years can make it hard to 'wean' children off these fiction-based strategies later on. They may mistakenly try to apply these 'fiction rules' to non-fiction:

- They will find it hard to use contents pages and 'skim through', preferring to start at the beginning of a non-fiction book.
- Their unfamiliarity with non-fiction will hold them back when they move up the school and are required to read textbooks, history books, maps, etc.
- They will also find it hard to write in a non-fiction style, like writing a report, and will lapse into more familiar story-style writing.
- They will find it hard to find the books they need for research.
- They will have a tendency to simply copy out verbatim from books rather than put things in their own words.

A broad range of fiction and non-fiction books is essential for teaching children how to read, from the moment they enter school.

What are non-fiction skills?

In the past, some schools have limited their teaching of non-fiction skills to a few brief lessons on 'library skills'. The potential of non-fiction instruction extends far beyond simply knowing about the contents and index page. The development of information skills is a complete literacy process. It includes:

- the ability to devise a purpose for reading (to find something out, to make something, etc.),
- the ability to select an appropriate book,
- the ability to use different reading strategies (skimming, scanning etc.),
- the ability to read different non-fiction structural guiders (headings, index, etc.),
- understanding how to gain information from photographs, charts and diagrams (visual literacy),
- understanding how to use different text types (an encyclopedia, a guide book, etc.),
- the ability to organise and present any findings from this research.

This process of purposeful reading involves speaking and listening, reading and writing (see diagram below). *Discovery World* has used this key process in developing its skills system to ensure children are taught all the key non-fiction strategies.

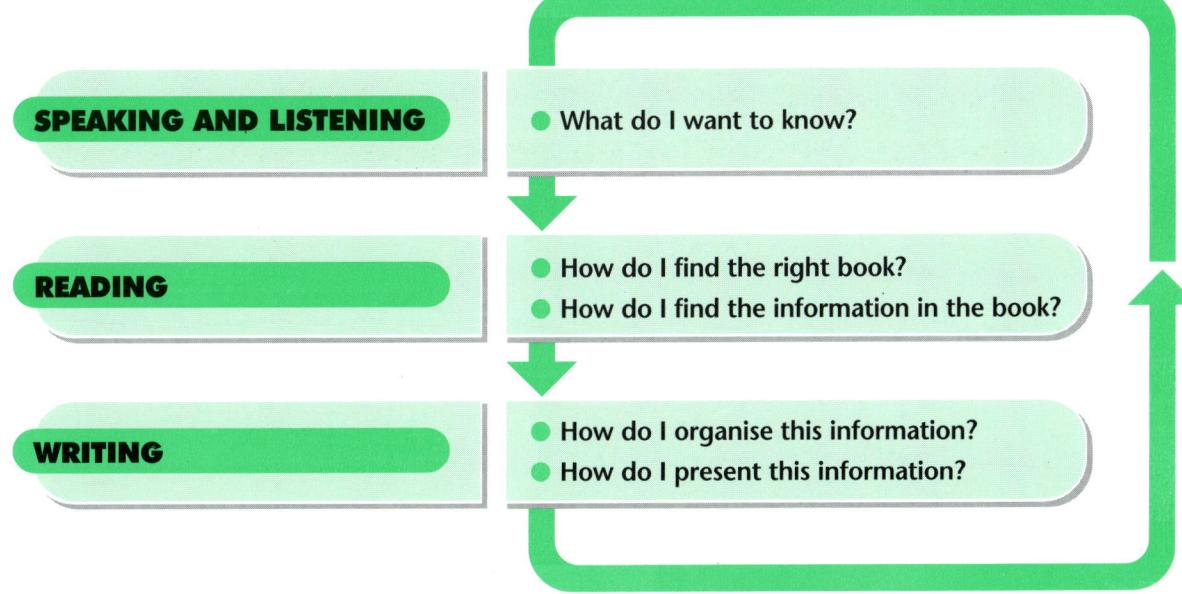

SPEAKING AND LISTENING
- What do I want to know?

READING
- How do I find the right book?
- How do I find the information in the book?

WRITING
- How do I organise this information?
- How do I present this information?

The information retrieval process

To make the skills involved in the information process easier to teach we have divided them into four main areas.

1 Reading and using the different parts of a non-fiction book

Every book has features which help readers understand the author's message. We call these features 'structural guiders'. Non-fiction books have far more structural guiders than fiction. They include front covers, contents pages, headings, sub-headings, labels, captions, glossaries, etc. If young children do not understand the purpose of these structural guiders they will not be able to use non-fiction books effectively. The *Discovery World* skills system is based on the premise that we cannot assume children simply 'absorb' these kinds of skills: they need to be taught.

For example, children need to be taught how to 'read' a non-fiction book cover. They need to understand how to interpret the photo or illustration on the cover, as well as the book title. Without lessons on front covers young children may assume *The Frog Prince* is an information book about frogs. When they are older they may have great difficulty selecting appropriate books for different topics or themes.

Many children spend quiet reading time poring over a non-fiction book. As teachers we have to ask ourselves how much those children really understand about how the book works. Do they understand why some words are picked out and placed next to illustrations (labels)? Do they know what order to read a non-fiction spread in? Do they realise how helpful the headings can be? It is no wonder that some of those non-fiction readers are in fact just looking at the photographs and avoiding the text. They need explicit instruction to show them how useful all these structural guiders can be.

For further information see page 27

2 Essential non-fiction reading skills

One of the most important non-fiction strategies readers need to develop is to clarify their purpose for reading – what do they want to find out? This involves the ability to ask relevant questions: a fundamental skill.

The readers then have to select books and decide how to read the book. With non-fiction this can include:

- *dipping into* the book, opening the book anywhere just for pleasure,

- *flicking through* the book quickly to get a general impression of what the book is about (skimming), or to look for a specific piece of information (scanning),

- *close reading* of a page, focusing on understanding what the author has written.

All these different reading strategies need to be taught. The **Discovery World** *Literacy Lesson Books* contain strategies to lay the foundations of these skills with even the youngest of children.

For further information see page 28

3 Reading and creating charts and diagrams

Non-fiction books use charts and diagrams (graphic organisers) to present information in accessible and usable ways. Charts and diagrams showing life cycles, timelines, production processes, size comparisons, maps, etc. occur in **Discovery World** books covering many areas of the curriculum. Readers need to be taught how and why these sequences and structures organise the information they present. Developing this 'visual literacy' also enables children to draw their own charts and diagrams. This is one of the earliest ways children can demonstrate taking notes. Instead of copying out text from a book, they are processing the information and presenting it in a new form. It is therefore a marvellous way to assess whether a child has really understood a process (e.g. if they mistakenly draw a butterfly turning into a caterpillar!).

For further information see page 29

4 Reading and writing different types of text

In order to write in different text types or genres, children need to be taught how these genres work in published texts. Non-fiction genres of recount, report, procedure, explanation and argument are all represented within *Discovery World*. Other forms introduced and taught are diary, biography, guide book, encyclopedia, dictionary, etc. Understanding how to read and write these text types is a long process, but any learning must be built on prior understanding of how authors write in different styles for different audiences. For example, it is important to understand that to follow a recipe one must first assemble the list of ingredients. Understanding that the recipe follows numbered steps is also essential. As children progress through the school the *Discovery World* programme will help you dig deeper and deeper into this knowledge of genre. For example, you can point out to older children that recipe instructions usually begin with a verb (mix the … , add the … etc.).

For further information see page **29**

Programme chart

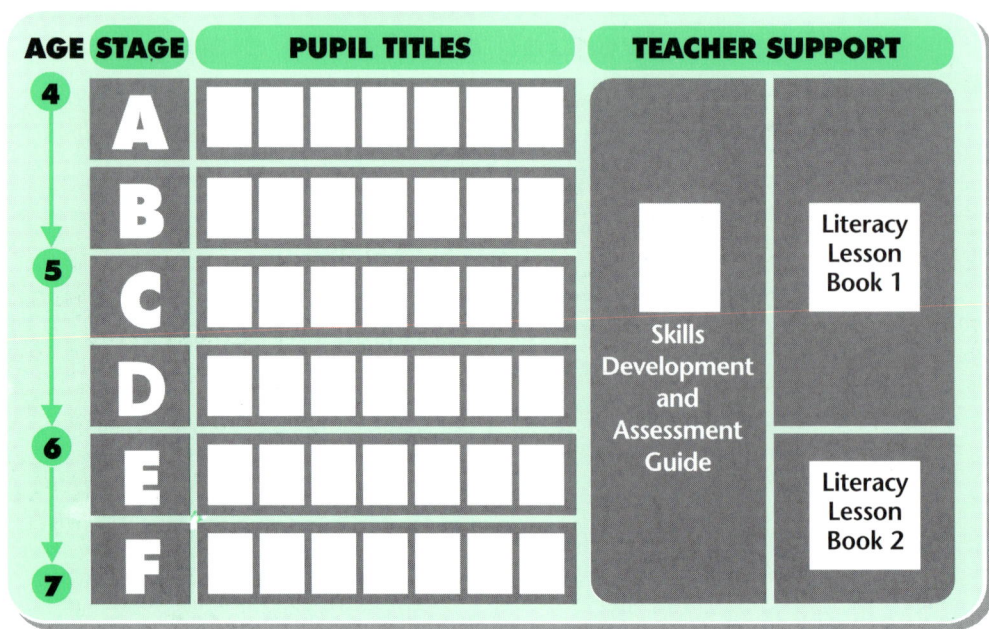

AGE	STAGE	PUPIL TITLES	TEACHER SUPPORT	
4	A			Literacy Lesson Book 1
5	B		Skills Development and Assessment Guide	Literacy Lesson Book 1
5	C		Skills Development and Assessment Guide	
6	D		Skills Development and Assessment Guide	
6	E		Skills Development and Assessment Guide	Literacy Lesson Book 2
7	F			Literacy Lesson Book 2

Components explained

The Literacy Lesson Books

The *Literacy Lesson Books* are the heart of the **Discovery World** skills system. They are designed to be used as a planned teaching programme for non-fiction reading and writing strategies throughout the school. The *Literacy Lesson Books* can be placed on the lap or on a table for group teaching. They contain a series of lessons, each with a specific learning objective. Each lesson has notes for the teacher and enlarged pages for the children to look at to focus discussion. Differentiated follow-up activities ensure children reinforce the lesson through practical activities. There are also linked photocopy masters (found in this book).

For further information see page 32

The Pupil Books

Discovery World has 42 systematically graded information books for the age 4–7 range. Many of these books will also be suitable for older children. They progressively introduce a wide variety of non-fiction features (structural guiders) and include excellent models of different text forms or genres. They cover a range of topics suitable for use across the curriculum, based on extensive research into teachers' needs. Every book is based on a different design, so that children have access to a wide variety of presentations.

For further information see page **37**

The Skills Development and Assessment Guide

This is designed for easy access to the information required for planning, teaching and assessing children's non-fiction skill development. The dip-in format supports the busy teacher, providing sensible charts to help make links between skills, books and topics. The photocopy masters highlight the key skill being practised and key features for assessment.

Big Books

Large format versions are available for some of the pupil books. These have been selected as the best examples to model and teach reading skills focusing on one book. They can be used to reinforce and follow up the skills introduced in the *Literacy Lesson Books*. Since the literacy lessons in **Discovery World** are based on big book techniques, we hope that you will find it increasingly easy to connect the two when teaching.

How to use *Discovery World*

TO HELP TEACH TOPICS

- Select *Discovery World* books from several stages to use in a topic box.
- Introduce features such as glossaries to help children develop topic vocabulary.
- Use *Discovery World* books as models for making books on chosen topics (e.g. a topic dictionary).

See page 42.

AS PART OF YOUR LITERACY TEACHING

- Use *Discovery World* pupil books as part of the graded reading resources in the classroom.
- Choose selected pupil books for individual take-home reading.
- Choose selected pupil books for focused group reading.
- Use *Discovery World* big books for shared reading (children rarely hear non-fiction read aloud and will benefit from you modelling expression and intonation when reading non-fiction).
- Use *Discovery World* in conjunction with other literacy programmes:
  - See links with *Storyworlds* and *Rhyme World* (page 19).
  - See links with other programmes (page 20).

TO HELP TEACH SUBJECTS

- Make copies of pages 45–47 and pass to relevant subject co-ordinators.

TO HELP TEACH NON-FICTION SKILLS

1 Choose a skill area from the *Literacy Lesson Book* contents page or from pages 30–31.

You may wish to introduce a new skill or to revisit a previously taught skill at a higher level.

2 Teach the skill to a group using the *Literacy Lesson Book*. Through discussion and example (modelling) the children see you (the skilled reader) using non-fiction effectively.

3 Follow up with group, reinforcement and extension activities.

Now the children can practise what they have learned (the lesson objectives will only be achieved if the children are involved in a practical activity).

4 Use the follow-up photocopy masters.

5 Complete assessment sheets (see pages 49–52). These sheets provide data which demonstrate children's ability to apply particular non-fiction skills when reading and writing.

Non-fiction and the school reading and language policy

Read 'Using *Discovery World* as a skills system' (page 26 onwards) to help you judge how much your school (and the school's reading and language policy) addresses the issue of non-fiction skills. One of the simplest ways to do this is to share the *Literacy Lesson Books* contents pages with other staff. Using these contents pages will help you to ensure that skills development is planned across a term by all teachers. They will also help you to plan a non-fiction skills development programme as part of your reading and language school policy.

Agree what terms to use

It may be helpful to agree terminology that all teachers will use across the school connected to non-fiction (e.g. will you use the term 'information book', 'fact book' or 'non-fiction book'? Will you teach the children to 'flick through', to locate 'key words'?).

Consider your topic planning

You can use the 'Links' section in this *Guide* to help you link topic plans to your non-fiction skills development. This will be useful when making planning decisions about writing expectations and assessment of topic understanding.

Conduct a school non-fiction audit

- Do you have enough non-fiction books readily available in the classroom?
- Are they logically organised?
- Do all teachers send non-fiction books home for reading with parents?
- Is the school library fully resourced and clearly organised?

Discuss the state of the current information books in the school

- Are they attractive and stimulating for the children?
- Are they up-to-date?
- Do they represent a variety of non-fiction genres?
- Do they have contents and index pages?
- Are there enough books written at a level young children can read?
- Do they resource the breadth of topics you cover?
- Do they contain examples of different kinds of structural guiders?

Organising *Discovery World* resources in the classroom

Which books do I need and where do I store them?

For the teacher

As the *Literacy Lesson Books* are the heart of the skills system we strongly recommend that these are *not* stored centrally or in the staffroom. Each teacher will need a copy readily accessible in the classroom. A Reception teacher would only need Book 1. Other teachers will need both books to accommodate a wide ability range. In this way the two *Literacy Lesson Books* can be used as a systematic programme, but also opportunistically when needed to support children's working knowledge of a skill.

We recommend you set aside an 'information skills' section on your shelves for the *Literacy Lesson Books,* and do not allow the children free access to them (they are a teacher-directed resource). We recommend you place a copy of this *Guide* with the *Literacy Lesson Books.* It is also very useful to have a set of **Discovery World** pupil books on your shelves so you can pick up the examples mentioned in the literacy lessons. (You could put other good examples here that would otherwise be lost amongst general reading materials. The **Magic Bean** In Fact series is a good example of excellent texts that can be set aside and used for skill instruction.)

For the children

- *Sort according to reading level.* Divide up books in the book boxes according to graded level. See 'links to other programmes' chart on page 20.

- *Sort according to topic.* Use topic links chart on pages 42–43. We recommend you buy group sets of the most relevant titles for your year's topics. This enables you to teach skills and use guided or shared reading strategies within a relevant topic area.

- *Sort according to genre.* You may decide to have a special non-fiction section in your classroom. Other reference books could also be placed here.

This chart is designed to help you make informed choices about using *Discovery World* in your school, catering for a wide ability range.

DISCOVERY WORLD RESOURCES – WHAT DO YOU NEED?

AGE	RESOURCES FOR THE TEACHER		RESOURCES FOR THE CHILDREN	
	Essential	Optional	*Essential*	Optional
4 ↓ 5	• Literacy Lesson Book 1 • Skills Development & Assessment Guide	• Big books from stages A & B • Stage A & B pupil books (accessible only to teacher to use as examples)	• Stage A & B pupil books	• Stage C & D pupil books • Group sets of titles for group/guided reading
5 ↓ 6	• Literacy Lesson Book 1 • Literacy Lesson Book 2 • Skills Development & Assessment Guide	• Big books from stages A – D • Stage A – D pupil books (accessible only to teacher to use as examples)	• Stage A – D pupil books	• Stage E & F pupil books • Group sets of titles for group/guided reading
6 ↓ 7	• Literacy Lesson Book 1 • Literacy Lesson Book 2 • Skills Development & Assessment Guide	• Big books from stages C – F • Stage C – F pupil books (accessible only to teacher to use as examples)	• Stage C – F pupil books	• Stage A & B pupil books • Group sets of titles for group/guided reading
7	See *Discovery World* age 7+ stages. Due to the wide ability range at this level many children will benefit from literacy lessons in *Literacy Lesson Book 2* – and enjoy using pupil books from stages E and F. For reluctant readers, pupil books from stages A onwards are ideal. Their real-world context will motivate the slower reader who does not want the stigma of using very early 'readers'.			

How do I store big books?

If you decide to buy the *Discovery World* big books we recommend you take special precautions when storing them. Big books can easily get torn or damaged so you may like to:

● Buy large clear envelopes specially designed for big book storage from specialist suppliers.

● Use spring-clip trouser hangers to hang up big books on hooks or rails.

● Punch a hole in the top left-hand corner and tie a loop of string through it. Big books can then be hung up on a hook.

Links

Links to *Storyworlds* and *Rhyme World*

Discovery World can be used alongside any other literacy programme. However, special attention has been given to links with its two 'sister' programmes: *Storyworlds* and *Rhyme World*.

Storyworlds is designed to teach reading through meaningful stories based on a careful progressive **key word** structure.

Rhyme World is designed to teach literacy using a range of materials designed to develop children's phonological awareness and ability to de-code using **phonics**.

All three 'sister worlds' combine to give children a balance of different strategies and skills to create motivated, successful readers. Each 'world':

● is carefully graded,

● has exceptional support for the beginning reader,

● covers a broad range of styles (story styles, non-fiction styles, styles of rhyme),

● has large-format books and materials with teacher activities and prompts to make them easy to use and accessible to a group.

This chart has been designed to help you run all three programmes together.

AGE	DISCOVERY WORLD STAGES	STORYWORLDS STAGES	RHYME WORLD STAGES
4	A	1	1
	B	2	
		3	2
5	C	4	
		5	3
	D	6	
6	E	7	
		8	4
7	F	9	

Links to other reading programmes

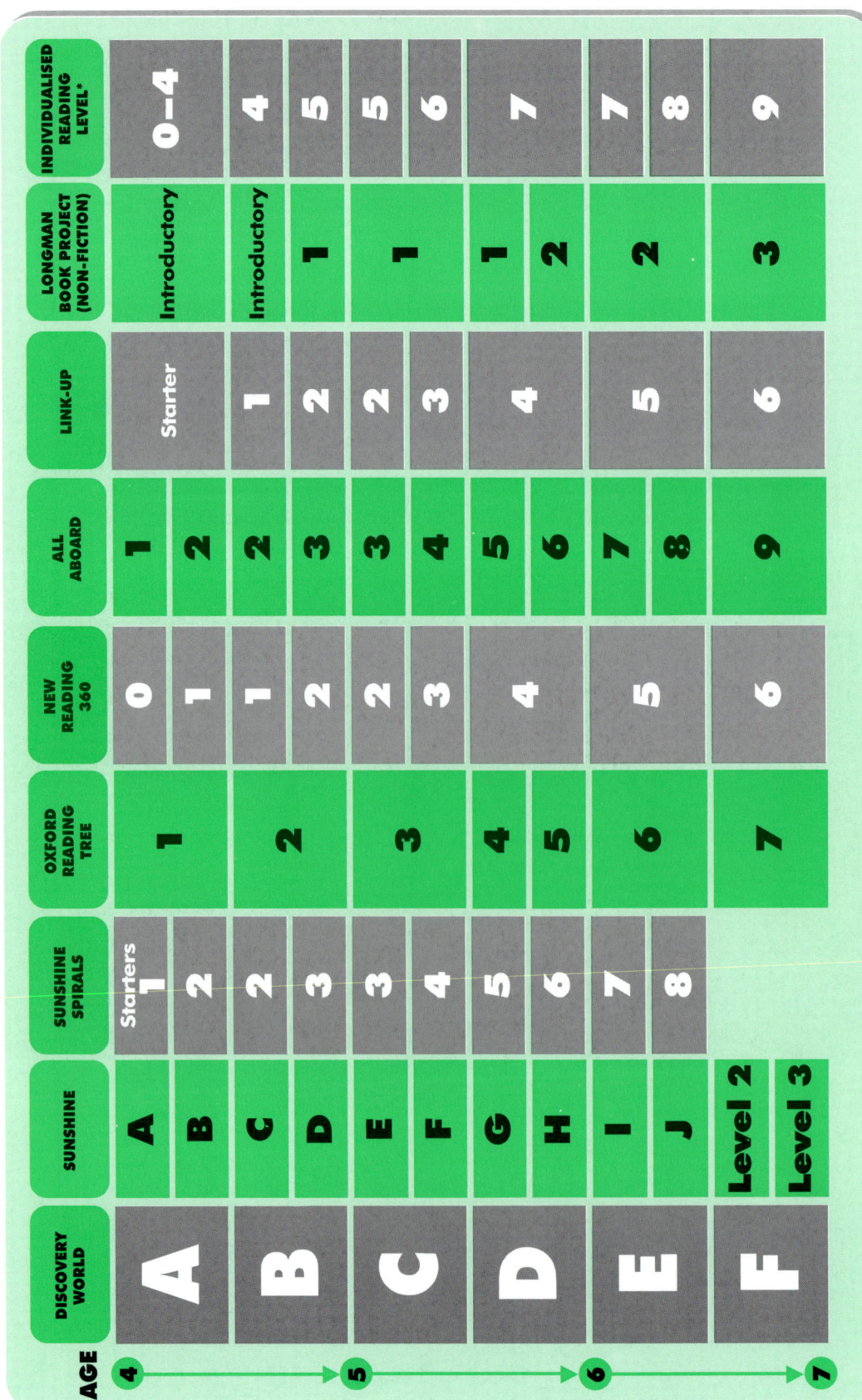

* The Individualised Reading Levels are compiled by Cliff Moon.

Links to the National Literacy Project

The National Literacy Project has already been heralded as one of the most useful frameworks for literacy teaching ever produced in this country. The framework's objectives are set out in termly units to provide a structure of progression and cover the required range of work in fiction and non-fiction reading and writing. Each term's work is focused on particular genres of reading and related purposes for writing.

There are three strands to the framework:

- **text level**   comprehension and composition
- **sentence level**   grammar and punctuation
- **word level**   phonics, spelling and vocabulary

Discovery World – at a text level

The *Literacy Lesson Books* are designed primarily to focus on non-fiction at a text level. The photocopy masters are also designed to develop comprehension and composition strategies. In order to look at books at a text level you need to focus a *group* of children's attention on the same text. *Discovery World* provides a number of ways to support group teaching strategies:

- using the *Literacy Lesson Books*,
- using *Discovery World* big books,
- using group sets of the pupil books,
- using copies of the photocopy masters.

The linking of reading and writing has also been emphasised within the National Literacy Project. *Discovery World* presents a range of text types/genres which enable children to focus on audience and purpose. This will directly influence their ability to write in different forms and for different purposes across the curriculum.

You can find a complete list of text types covered in *Discovery World* on page 38.

Discovery World – sentence level

Non-fiction books use language in special ways and thus provide an important context for the study of language. Developing knowledge about the ways in which grammar, punctuation and vocabulary are used in information text is important for both reading and writing effectively.

Discovery World – word level

Each *Discovery World* pupil book has a particular subject focus. You can use this focus to look at specialist vocabulary relevant to that subject. Literacy lessons on using the index or glossary will enable you to extend this study by assembling your own word banks and glossaries etc. One centrally important focus for many of the literacy lessons is developing the idea of a 'key word' (for example, the word 'insect' is particularly important when studying 'minibeasts'). Key word study will help children develop a wider, more subject-appropriate vocabulary.

Links to the National Curricula

Although the following references focus mostly on reading, it is important to recognise the vital part information skills play in the development of children's *writing*. Factual writing is organised in different ways. Only by reading and discussing good models of different types of non-fiction texts will progress be seen in children's writing. All national curricula and guidelines recognise the importance of a sense of audience and form.

Links to English in the National Curriculum (England and Wales)

The revised Orders for English outline the importance of experiencing a wide range of texts from school entry. *Discovery World* is the first non-fiction programme that provides truly age-appropriate lessons to introduce non-fiction to very young children.

The list below highlights the key areas *Discovery World* covers.

Key Stage 1 Reading

Range

- read from a range of genres
- use reference materials (including dictionaries and encyclopedias)
- read and discuss content
- use a variety of organisational and presentational features

Key Skills

- use reference materials for different purposes
- learn about the structural devices for organising information (e.g. contents, headings, captions)
- use visual information to support understanding before, during and after reading
- use rereading to make sense of unfamiliar text
- develop response to books
- develop word banks based on spelling and sound patterns
- identify key vocabulary
- identify critical features of non-fiction language

Standard English and Language Study

- consider the characteristics and features of different kinds of texts

Links to English in the Northern Ireland Curriculum

The list below highlights the key areas *Discovery World* covers at Stage 1:

Progression in reading skills is outlined in *Discovery World* through focused teaching using the *Literacy Lesson Books* and differentiated reading activities using a wide range of books.

Contribution to education (cross-curricular) themes is developed by the use of reading to explore topics from all curriculum areas and to teach the appropriate strategies and forms for each area.

Contexts for teaching and learning include whole class, group, paired and individual activities.

The **range of information texts** has been considered to support children at this early stage. Each book has been designed to support young readers and to provide essential information about a topic. Many ways of organising and presenting information are used, especially high-quality illustrations and photographs.

EXPECTED OUTCOMES	DISCOVERY WORLD
A make choices for themselves	Include *Discovery World* books as part of your classroom collection for individual, shared and take home reading.
B build a sight vocabulary	Banks of key topic words can be built from the pupil books.
C use a range of strategies	Use *Discovery World* Literacy Lesson Books.
D use evidence from text	*Discovery World* reading activities encourage children to use visual and text information when predicting, inferring and deducing.
E talk about written language	Literacy lessons include teaching about words and sentences.
F recognise spellings	Developing 'key word' banks will assist personal and cross-curricular writing.
G respond to texts	*Discovery World* reading activities include response to reading about a wide range of topics. Response to visual features of texts are also developed, e.g. a life cycle sequence.
H make use of dictionaries, word banks, contents and indexes	All titles include a contents and index. There are specific literacy lessons that focus on these text types and features.
I collect information from many sources	*Discovery World* activities encourage children to collect and organise information in different ways.
J read aloud	It is important to use non-fiction books for read aloud sessions by both teachers and pupils, to encourage development of flexible reading expression and intonation.
K read independently	Titles cover a wide range of topics and styles.

Links to English Language 5–14 (Scottish National Guidelines)

English Language 5–14 Programmes of Study for Reading recognise the importance of reading for information from Level A. *Discovery World* books provide a rich resource for reading for enjoyment too. The list below highlights the key areas *Discovery World* covers.

LEVEL

READING FOR INFORMATION

A
- simplest information, signs, labels, notices
- setting questions for reading

B
- environmental texts, instructions, letters, word banks
- using the index, contents
- predicting from cover, using questions

LEVEL

READING FOR ENJOYMENT

A
- regular planned time for self-chosen reading
- model good reading habits

B
- discussion related to reading and response

LEVEL

RESPONDING TO REFLECT ON THE WRITER'S IDEAS AND CRAFT

A
- discussion related to reading and response to information presented

B
- encourage prediction
- question texts, during and after reading
- attend to picture and context clues
- recall information and link to own experience

LEVEL

AWARENESS OF GENRE OR FORM

A
- recognise differences of genre or form between fiction and non-fiction texts

B
- identify features of text including page layout, content organisation, visual information

LEVEL

READING ALOUD

A
- read aloud favourite titles

B
- read fluently and with appropriate expression

LEVEL

KNOWLEDGE ABOUT LANGUAGE

A
- identify words, sentences, numbers used in a variety of ways, e.g. labels

B
- identify parts of a book and page and explain their purpose

How to use Discovery World as a skills system

Quick contents

What skills to teach?

The skills of a proficient reader are transparent and usually carried out with such efficiency that it is very difficult to separate out the many skills the proficient reader is using. It is important to make explicit for children what these skills are, and one of the most effective ways to do this is through 'modelling'. Modelling simply means acting out what skilled readers do, explaining the thinking processes and skills being used for the novice reader. The *Discovery World* skills system is based on a comprehensive list of non-fiction skills which children need to use as readers and writers of information texts. It is important that these skills are developed as part of literacy understanding as well as part of the information process.

In *Discovery World* we have grouped these skills into four areas for focused study:

- Reading and using the different parts of a non-fiction book.
- Essential non-fiction reading skills.
- Reading and creating charts and diagrams.
- Reading and writing different types of text.

Reading and using the different parts of a non-fiction book

All books have common features which readers use to understand the context and content of the book. Book covers, for example, are an important starting point for focused reading and can help readers both prepare for the task ahead and begin to consider what information the book might contain. The title, cover illustration and back cover blurb all provide useful clues to the content of the book. They can also be misleading sometimes. It is often taken for granted that young children understand why covers are organised and presented in particular ways, and it is important that the decisions made by an author, designer or publisher are seen as deliberate rather than haphazard or magic.

Non-fiction books have far more complex and diverse page lay-outs than fiction. There are different headings, different type-faces, captions, labels, etc. It cannot be assumed that children know how and why each of these are used. Labels often seem to be words 'floating on a page' and readers can be unsure exactly what is being labelled – the whole or part of an illustration. Understanding the different purposes for labels, captions, head-ings and sub-headings must be seen as skills which require teaching so that novice readers are empowered in selecting important information.

Contents, indexes and glossaries are structural guiders which most non-fiction books use, and many novice readers have little knowledge of why they are used or how they are compiled. The teaching of non-fiction skills has often focused on the use of these features, but it is not always explained *why* they are useful. (Research has shown that even readers who understand how to use these features seldom do so in practice.) It is important that children realise that looking at the contents is a good way to find out what a book is about. An index is a better place to look if the reader wants to look for specific information.

Essential non-fiction reading skills

The special features and skills used in reading non-fiction need to be modelled and taught explicitly to young readers. At the earliest levels readers need to understand that words, illustra-tions and photographs convey information and are interrelated. They need to learn the difference between fiction and non-fiction, something many readers will need plenty of practice at before they understand the concept. Children need to learn how to ask questions about a subject, and how to list what they already know and what they want to find out. They need to learn how to select a book that may help them find the answers to these questions. They need to know how to flick through a book looking for something specific (skimming and scanning). They need to learn how to search for 'key words' in a book, look-ing first at headings or labels before they read the text. Once they have found something out, children need to learn how to record and organise what has been read in written outcomes or by creating charts or diagrams.

Reading and creating charts and diagrams

Non-fiction books use a variety of charts and diagrams (graphic organisers) to present information in visual ways. Many of these make reading and comparing information much easier. Once they have understood these organisers, children can make their own charts and diagrams, which is an early form of note-taking (taking information and condensing it in a new and easy-to-follow way).

Charts condense information in two-way columns. These are used widely in the environment and need to be read in specific ways. You will need to model for the children how to read them across the page or up and down columns.

Diagrams are used to represent the actual view of something in a clearer way. For example, the use of a simple diagram next to a photograph helps readers see detail more clearly. Diagrams can also show how processes work, one of the most important being the movement of time in a timeline. Often diagrams show information visually more simply than if the same information had been written. Graphs, scales and keys use mathematical references like centimetres and will need explanation.

Reading and writing different types of text

All writers use text structures or genres to convey information to a particular audience. These structures are not arbitrary and enable readers to access information in predictable ways. For example, recipes written in a prose format are difficult to read while cooking – the 'procedural genre' is more easily read (e.g. list ingredients needed, number the next steps to be followed, etc.).

Readers need to understand that:

- different text types have different styles or layouts: for example dictionaries are in alphabetical order and have guide letters or words at the top, diaries usually have a date at the top, etc.,
- they can copy the 'rules' of a particular genre and write their own diary or recipe, etc.

The chart on the next page sets out all these skills. It indicates which literacy lesson covers the particular skill, and gives an approximate indication of the age group of the children for which the lessons are designed. It also indicates which photocopy masters support the teaching of each skill.

Skills Chart

This skills chart sets out all the learning objectives covered by *Discovery World*, in four main areas. You can use this to plan which literacy lesson you want to teach. We have added suggested age correlation, but of course this will depend on the ability of your pupils and the pace you want to go.

Lesson number in **bold** = main lesson.

Lesson numbers in roman = also covered in this lesson.

TEACHING OBJECTIVE	LITERACY LESSONS		
	AGE 4–5	AGE 5–6	AGE 6–7
READING AND USING THE DIFFERENT PARTS OF A NON-FICTION BOOK			
Understanding front covers	4	**6** ◦ 7 ◦ 11	17 ◦ 27 ◦ 29
Understanding what a 'double page spread' is	5		
Understanding page numbers		8 ◦ 13	
Understanding page layout		8	
Understanding labels and captions		8 ◦ 14 ◦ 15	21 ◦ 22 ◦ 25
Understanding page headings/sub-headings		8 ◦ 13 ◦ 14	
Using the contents page		**13**	**19** ◦ 20
Using the index			19 ◦ **20** ◦ 26
Using back cover blurbs		6 ◦ 17	
Using a glossary			**26**
ESSENTIAL NON-FICTION READING SKILLS			
We get information from photographs (and pictures)	**1**		
We get information from print	**2**		
We get information from books	**3**		
The difference between fiction and non-fiction	**4**	6 ◦ 9	17 ◦ 27 ◦ 29
How a non-fiction page (book) works	**5**	**8** ◦ 13	
Non-fiction books are about different subjects		**7**	
How to 'dip in' or 'flick through' (early skimming & scanning)		**9** ◦ 12	17 ◦ 19
How to ask questions		**10** ◦ 11	18 ◦ 28
Identifying what you want to find out		**11**	
Looking for key words		**12** ◦ 13	17 ◦ 19 ◦ 20 23 ◦ 24 ◦ 26
Choosing the best book			**17**
Non-fiction books have different purposes	4	9	**18** ◦ 27
Using a library		7	**27**
Distinguishing between fact and opinion			18 ◦ 19
Understanding alphabetical order			20 ◦ 23
Early note-taking			30

TEACHING OBJECTIVE	LITERACY LESSONS		
	AGE 4–5	AGE 5–6	AGE 6–7
READING AND CREATING CHARTS AND DIAGRAMS			
Understanding simple charts		**14**	30
Understanding simple diagrams		**15**	21 · 22
Reading and making diagrams			**21 · 22**
Scale diagrams/comparisons			21 · **22**
Diagrams/maps with keys			**22**
Number/colour keys			**22**
Timelines			18
Simple maps/street plans			18
Aerial maps/route maps			18 · 21
3-D maps			18 · 21
Cut aways/cross-section diagrams			21 · 22
Family trees			21
Bar charts			21
READING AND WRITING DIFFERENT TYPES OF TEXT			
Simple environmental print	2		
Signs and notices	2		
Lists of instructions	3		
Atlas	4		**18**
'How to …'/instructional text		7	**18**
Procedural text/recipes	4	7	**18**
Biography		10	**18**
Autobiography		10	18
Interview			18
Dictionary			**23** · 24
Encyclopedia		7 · 9 · 12	**24** · 27
Diary			**25**
Report on a visit/trip			25

How to use the *Literacy Lesson Books*

The *Literacy Lesson Books* form the heart of the **Discovery World** skills system and can be used in three ways:

1 Follow the lessons in the given sequence, from 1 to 30, systematically across the age 4–7 range.

2 Follow your own sequence, using the chart on pages 30–31 as a guide.

3 Simply choose lessons as and when opportunities arise in the classroom.

Whatever you do, you will need to constantly revisit lessons with certain children who need extra support and reinforcement. Skills can only develop over time – from introduction, through practice and onto application.

These lessons can take as little as 10 minutes or be a more in-depth 30 minute session, depending on the amount of interaction encouraged and children's familiarity with the learning objective and its uses.

Literacy Lesson Book 1

Lessons 1 – 5 are designed approximately for 4 – 5 year olds.

Lessons 6 – 15 are designed approximately for 5 – 6 year olds.

These lessons focus mostly on reading skills.

Literacy Lesson Book 2

Lessons 16 – 30 are designed approximately for 6 – 7 year olds.

These lessons focus on reading and writing skills.

Lesson 16 is designed to assess the children's current abilities.

Lessons 28 – 30 are designed to revisit all areas in a complete non-fiction mini-topic.

HOW TO USE THE *LITERACY LESSON BOOKS*

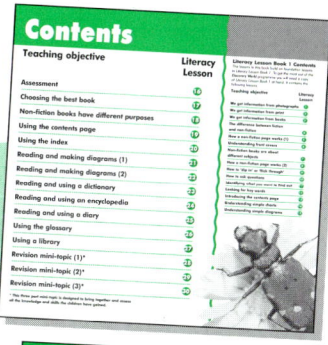

1 Select the skill/teaching objective you want to teach from the contents page or from pages 30–31 of this *Guide*.

2 Check the 'You will need' box for any resources required.

3 Place the *Literacy Lesson Book* on a table or your lap where the whole class or a group can see the pupil page.

4 Use the teacher's page for prompts and ideas to help you:
a) model how a competent reader uses non-fiction,
b) as a focus for shared reading and discussion about the text/skill.

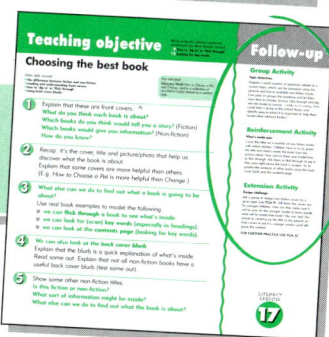

5 Select an activity from the 'Follow-up' section. Provide different levels of support for more and less able children.

6 Use the relevant photocopy master.

7 Fill in the assessment sheet on page 49 – plan to revisit this lesson, a previous lesson, or move on to the next lesson according to learning outcomes.

The *Literacy Lesson Books* are designed to be used in a simple teaching sequence using a basic four-step sequence:

Using the Literacy Lesson Book

1 Modelling   *Show me how to do it*

2 Shared reading   *Do it with me*

Using a Follow-up Activity

3 Supported action   *Watch me do it*

4 Independent action *Now I'll do it on my own*

1 Modelling: *Show me how to do it*

Modelling is simply a way to show children how a process or skill works in action.

Modelling is not isolated skill-based instruction, but rather a way of making explicit the actions readers take while reading. It is an explicit teaching strategy and an essential part of the learning process underpinning **Discovery World**. Put very simply, it is telling the children what you are doing as you read ("I want to use this non-fiction book. First I look at the front cover. This is the book title." Etc.).

All the modelling strategies used in the *Literacy Lesson Books* can be reinforced with similar sessions using big books. (Many of the **Discovery World** titles are available in big book format. *The Magic Bean* In Fact series also has an excellent reputation for providing quality big book texts.) Big books also enable you to repeat the same skill in the context of a different book. This will be helpful in supporting less able children and for reinforcement of particular skills.

When modelling remember to:

Read aloud

● Always read out the texts used in the *Literacy Lesson Books*. Share them with the children. Take every opportunity to also read aloud big books or other non-fiction texts. Let the children see how you use different book features like headings.

Talk aloud

● Tell children what you are doing as you read to them.

● Talk about the actions you take, like a series of steps ("I want to find out what this picture is about. This word next to it may help. It's called a label." Etc.).

Think aloud

● Tell children what you are thinking and deciding as you read to them.

● Talk about the predictions you make about content.

● Talk about the questions you ask about content.

● Talk about the page features you attend to, before, during and after reading.

● Talk about the links you make to knowledge outside the book.

● Talk about what you have understood during and after reading.

2 Shared reading: *Do it with me*

Invite the children to help you with the reading task when you are using the *Literacy Lesson Books*. This joint action with children ensures they take some responsibility for the process. Children's questions and decisions can guide your continuing action in completion of a non-fiction reading or writing task. The discussion during the task is more important than speedy completion.

Having modelled a process once (for example using a contents page), revisit the whole process using shared reading strategies:

● Encourage children to describe the next step (where will I find the contents page?).

● Ask children to give reasons why a particular step is useful or not.

● Ask children to use the skill with a book in front of the group.

● Talk about the purpose for using the skill with current topic books.

3 Supported action: *Watch me do it*

The follow-up activities in the *Literacy Lesson Books* provide the practice needed to use the skill in real ways. The 'group activity' is designed as a general follow-up, often for the whole class. You may choose to move some groups directly onto the differentiated 'reinforcement' or 'extension' activities according to ability. (The reinforcement activities are often designed for adult supervision.)

Again, the emphasis is not on isolated skills practice, but on using the skill for completion of non-fiction reading and writing tasks. Children can work as a class or in groups or pairs. Work with the group to define the task, reminding children of the use and purpose of the skill you have modelled, supporting children in beginning the task, clarifying the next steps, etc. Share completed tasks with the whole group or class in a plenary session. Discuss any problems that arose or lessons learned.

4 Independent reading: *Now I'll do it on my own*

Using the photocopy masters or setting an independent activity provides opportunities for children to apply the skill with minimum support. This is an ideal time to assess understanding and plan the next lesson.

You may choose to leave the *Literacy Lesson Book* open after completion of a lesson where the children can see it. At any point over the next few days you could ask individuals or pairs to retell the lesson as they understand it using the open page.

Reading and writing different types of text

Discovery World includes examples of the main non-fiction genres. Many books draw on more than one genre or form in exploring content, as happens in real life. For the age 4–7 range we have limited specific text type literacy lessons to the most obvious and useful non-fiction examples: dictionaries, encyclopedias, diaries, atlases, 'How to' books and biographies.

These, and other common genres and forms, are very supportive structures for use when writing. For example, when using *Korky Paul: Biography of an Illustrator* children can:

● try to write in the same genre about a different *subject* (e.g. a biography of a friend),

● try to write about the same subject in a different *genre* for a different purpose or audience (e.g. a newspaper article about Korky Paul).

A complete list of the different text types found in the age 4–7 *Discovery World* books can be found in the following chart.

DIFFERENT NON-FICTION TEXT TYPES IN *DISCOVERY WORLD*

Columns are grouped by Stage (A–F):

- **STAGE A:** Seasons · Which is Alive? · Day & Night Animals · Special Clothes · Shapes · Shopping · Choosing Cards
- **STAGE B:** Animal Legs · My Body · Just Add Water · Sizes · Homes · My History · What Did I Use?
- **STAGE C:** Minibeast Encyclopedia · My Bean Diary · Eyes · Materials · Using Tools · Breakfast · A Day in the Life of a Victorian Child
- **STAGE D:** Amazing Eggs · Change · Looking at Light · Fun Things to Make and Do · Time for a Party · A Closer Look at Parks · What's Underneath?
- **STAGE E:** How to Choose a Pet · Insect Body Parts · Animal Movement · Everyday Forces · Maps · Encyclopedia of Life in the 1950s and 1960s · Korky Paul: Biography of an Illustrator
- **STAGE F:** Prehistoric Record Breakers · Keeping Tadpoles (Alive!) · Science Dictionary · Materials: their Properties and Uses · My Holiday Diary · Transport Timelines · Arts and Crafts Around the World

MAIN GENRES

Genre	Titles with ✔
Recount	My History · A Day in the Life of a Victorian Child · A Closer Look at Parks · Transport Timelines
Description	Seasons · Which is Alive? · Day & Night Animals · Special Clothes · Shapes · Animal Legs · My Body · Sizes · Homes · Time for a Party · Animal Movement · Everyday Forces · Arts and Crafts Around the World
Procedure	Just Add Water · Fun Things to Make and Do · How to Choose a Pet · My Holiday Diary
Report	Eyes · Materials · Amazing Eggs · How to Choose a Pet · Materials: their Properties and Uses · Transport Timelines
Explanation	Breakfast · Change · Looking at Light · Everyday Forces
Reference	Minibeast Encyclopedia · How to Choose a Pet · Insect Body Parts · Encyclopedia of Life in the 1950s and 1960s · Prehistoric Record Breakers · Science Dictionary · Materials: their Properties and Uses

COMMON FORMS

Form	Titles with ✔
Simple environmental print	Day & Night Animals · Shapes · Shopping · Encyclopedia of Life in the 1950s and 1960s · My Holiday Diary · Transport Timelines
Signs and notices	Day & Night Animals · Shapes · Encyclopedia of Life in the 1950s and 1960s · My Holiday Diary
Lists of instructions	Fun Things to Make and Do · How to Choose a Pet
Atlas/maps	A Closer Look at Parks · Maps · Transport Timelines
'How to…', instructional text	Fun Things to Make and Do · How to Choose a Pet · My Holiday Diary · Arts and Crafts Around the World
Procedural text/recipes	Just Add Water · Fun Things to Make and Do · Arts and Crafts Around the World
Biography	Korky Paul: Biography of an Illustrator
Invitation	Time for a Party
Dictionary	Science Dictionary
Encyclopedia	Minibeast Encyclopedia · Encyclopedia of Life in the 1950s and 1960s
Diary	My Bean Diary · My Holiday Diary
Report on a visit/trip	My Holiday Diary
Personal account	My Holiday Diary
Postcard	My Holiday Diary
Travel brochure	My Holiday Diary
Menu	My Holiday Diary
Question/answer	Which is Alive? · Shopping · Homes · What Did I Use? · Materials
Guide book	A Closer Look at Parks · Animal Movement · Transport Timelines · Arts and Crafts Around the World
Interview	Korky Paul: Biography of an Illustrator

Reading and creating charts and diagrams

The reading of visual information has been neglected in many teaching programmes. It has been taken for granted that children read pictures easily, but this is often at a superficial level only. Children need to be taught a range of visual literacy skills in relation to books and reading as well as those for the visual media of film, television and advertising.

Non-fiction books present information using a large range of visual representations. **Photographs** are being used more widely than illustrations in children's books and are believed to present more accurate information. **Illustrations** have often misrepresented information and been inaccurate in detail, often being drawn from other illustrations rather than direct observation. We hope that in developing *Discovery World* we have been rigorous in selecting photographs and preparing illustrations which can be used for teaching visual literacy skills.

Non-fiction books also present information in other visual forms. **Diagrams** are an important visual representation of knowledge, and ways to read them can be discussed with young children. Attention to labelling and captions at later stages supports the introduction of specific vocabulary. Even simple diagrams can provide information which could not be presented easily in written text. Other diagrams use scale to show comparison to known objects or to measurements. **Cut-away and cross-section diagrams** show readers a view which cannot be achieved by photographs. It is important to talk to children about how the diagram has been drawn, how its accuracy is known and where the illustrator stood to get the view. This helps children understand the representative nature of diagrams.

Graphic organisers are another common visual form used in non-fiction books. Charts organise information so that it can be compared and contrasted in specific ways. For example, a **two column chart** is a simple comparison chart which young children often use when reading and writing information. A more complex **four column chart** compares more sets of information with one another.

Other charts such as **flow charts** show movement of time in processes or stages of change. **Picture sequences** can also show this change process, but sometimes they are not appropriate. Flow charts represent the pictures, with diagrams, captions or labels.

Tree charts show the relationship between pieces of information, these can be simple or complex and are more easily read when knowledge of the topic is at a high level.

Timelines summarise and organise information in time order. They present a lot of information in less space and, even more importantly, in one view.

Discovery World books use charts whenever they can provide information to enhance the written text. Often graphic organisers can replace wordy and difficult text – 'a picture is worth more than a 1000 words'. However, the main role of graphic organisers is to present information in a structured way to enable readers to understand more than the written text can explain.

Maps are another form of visual text which children experience early. The presentation of a view from above fascinates many readers, but reading them requires attention to a number of key features. These include orientation to known factors, scale of map, selection of details shown and not shown, use of symbols, etc. **Three dimensional maps** show a view, not from above, but to the side, an **aerial map** shows the view from directly above. **Plans** are one form of map and simple plans are the foundation for developing map reading and map making skills. **Keys or legends** show how symbols are used to represent information on a map. The use of colour is one feature, among many others, which need to be taught explicitly when working with maps.

The **typography** in non-fiction books uses letter size, shape and colour in particular ways. These can become important features for reading and understanding information as they are used to attract attention to significant parts of a page. **Labels, headings and captions** often utilise different print size from the main body of text. Other features need to be discussed as aspects of **page layout** which help readers in selecting which information to read for different purposes. Some examples of these are:

● boxes drawn around sets of information,

● use of background colour,

● markers such as bullet points,

● lines connecting information as in labelling and tree diagrams,

● arrows which show relationship or direction of movement.

Charts and diagrams as aids to writing

Many of the ways used to organise information in books are excellent note-making aids for recording what has been understood during reading. Recording charts encourage readers to select significant parts of the main text to record rather than copying out 'whole chunks of text'. They allow writers to:

- order information into a simple visual form,
- compare information gathered from different sources,
- clarify gaps in knowledge for further reading.

Information prepared in note form using some kind of diagram or chart will help children when they are preparing to write their own non-fiction books. Even a list helps writers know what to write in each sentence, or paragraph. A detailed timeline is an excellent organiser for writing a diary or log of a journey, life or change process.

CHARTS AND DIAGRAMS IN *DISCOVERY WORLD*

	STAGE A Seasons	Which is Alive?	Day & Night Animals	Special Clothes	Shapes	Shopping	Choosing Cards	**STAGE B** Animal Legs	My Body	Just Add Water	Sizes	Homes	My History	What Did I Use?	**STAGE C** Minibeast Encyclopedia	My Bean Diary	Eyes	Materials	Using Tools	Breakfast	A Day in the Life of a Victorian Child	**STAGE D** Amazing Eggs	Change	Looking at Light	Fun Things to Make and Do	Time for a Party	A Closer Look at Parks	What's Underneath?	**STAGE E** How to Choose a Pet	Insect Body Parts	Animal Movement	Everyday Forces	Maps	Encyclopedia of Life in the 1950s and 1960s	Korky Paul: Biography of an Illustrator	**STAGE F** Prehistoric Record Breakers	Keeping Tadpoles (Alive!)	Science Dictionary	Materials: their Properties and Uses	My Holiday Diary	Transport Timelines	Arts and Crafts Around the World
DIAGRAMS																																										
Simple									✔							✔						✔											✔			✔						
Scale diagrams/ comparisons											✔	✔			✔																					✔						
Cutaways/ cross-sections									✔								✔					✔					✔												✔			
CHARTS																																										
Column charts	✔	✔	✔						✔									✔	✔			✔							✔	✔	✔								✔	✔		✔
Bar chart																																									✔	
Flow charts									✔												✔	✔																✔	✔	✔		
Family tree																																		✔				✔				
Timelines																												✔						✔							✔	
MAPS																																										
Aerial maps																												✔					✔	✔							✔	✔
3D map																												✔					✔									
Number/colour keys																																	✔			✔			✔			

Using *Discovery World* to help teach topics

DISCOVERY WORLD TITLES

TOPIC/THEME	Seasons	Which is Alive?	Day & Night Animals	Special Clothes	Shapes	Shopping	Choosing Cards	Animal Legs	My Body	Just Add Water	Sizes	Homes	My History	What Did I Use?	Minibeast Encyclopedia	My Bean Diary	Eyes	Materials	Using Tools	Breakfast	A Day in the Life of a Victorian Child	Amazing Eggs	Change	Looking at Light	Fun Things to Make and Do	Time for a Party	A Closer Look at Parks	What's Underneath?	How to Choose a Pet	Insect Body Parts	Animal Movement	Everyday Forces	Maps	Encyclopedia of Life in the 1950s and 1960s	Korky Paul: Biography of an Illustrator	Prehistoric Record Breakers	Keeping Tadpoles (Alive!)	Science Dictionary	Materials: their Properties and Uses	My Holiday Diary	Transport Timelines	Arts and Crafts Around the World
	STAGE A							**STAGE B**							**STAGE C**						**STAGE D**								**STAGE E**							**STAGE F**						
Animals	✔	✔	✔					✔							✔		✔					✔	✔				✔	✔	✔	✔	✔					✔	✔	✔		✔		
Celebrations							✔																			✔								✔								
Change	✔									✔			✔		✔				✔			✔	✔		✔							✔		✔	✔		✔	✔	✔		✔	✔
Clothes	✔			✔																														✔								
Communication				✔			✔																											✔								
Dinosaurs																																				✔						
Electricity																								✔			✔							✔					✔			
Energy																																✔										
Environment	✔					✔						✔			✔								✔	✔					✔	✔		✔	✔			✔	✔	✔				
Families							✔						✔										✔			✔						✔	✔							✔		
Festivals							✔																			✔																✔
Food						✔				✔					✔					✔		✔	✔			✔	✔				✔						✔	✔		✔		
Forces																																✔							✔			
Growth	✔										✔					✔						✔	✔												✔			✔	✔			
Health									✔	✔																								✔								
History													✔								✔													✔	✔						✔	
Holidays																																			✔					✔		
Houses and homes												✔									✔						✔	✔				✔	✔									
Human body									✔								✔																						✔			
Journeys																																	✔							✔	✔	✔
Length											✔			✔																							✔					
Light & colour		✔												✔			✔						✔	✔										✔								✔

The content of the *Discovery World* pupil books was based on extensive research into teachers' most popular topics and themes. Non-fiction skills are best developed in the context of meaningful practical activities. By ensuring wide topic coverage we hope we will enable you to teach skills as part of your topic teaching. (You will notice that the 'follow-up activities' section in the *Literacy Lesson Books* often connects the learning objective to your current topic teaching.)

This chart will help you connect *Discovery World* titles to your term's topics and themes.

DISCOVERY WORLD TITLES

Titles are grouped by stage: **Stage A** (Seasons; Which is Alive?; Day & Night Animals; Special Clothes; Shapes; Shopping; Choosing Cards), **Stage B** (Animal Legs; My Body; Just Add Water; Sizes; Homes; My History; What Did I Use?), **Stage C** (Minibeast Encyclopedia; My Bean Diary; Eyes; Materials; Using Tools; Breakfast; A Day in the Life of a Victorian Child), **Stage D** (Amazing Eggs; Change; Looking at Light; Fun Things to Make and Do; Time for a Party; A Closer Look at Parks; What's Underneath?), **Stage E** (How to Choose a Pet; Insect Body Parts; Animal Movement; Everyday Forces; Maps; Encyclopedia of Life in the 1950s and 1960s; Korky Paul: Biography of an Illustrator), **Stage F** (Prehistoric Record Breakers; Keeping Tadpoles (Alive!); Science Dictionary; Materials: their Properties and Uses; My Holiday Diary; Transport Timelines; Arts and Crafts Around the World).

TOPIC/THEME	Seasons	Which is Alive?	Day & Night Animals	Special Clothes	Shapes	Shopping	Choosing Cards	Animal Legs	My Body	Just Add Water	Sizes	Homes	My History	What Did I Use?	Minibeast Encyclopedia	My Bean Diary	Eyes	Materials	Using Tools	Breakfast	A Day in the Life of a Victorian Child	Amazing Eggs	Change	Looking at Light	Fun Things to Make and Do	Time for a Party	A Closer Look at Parks	What's Underneath?	How to Choose a Pet	Insect Body Parts	Animal Movement	Everyday Forces	Maps	Encyclopedia of Life in the 1950s and 1960s	Korky Paul: Biography of an Illustrator	Prehistoric Record Breakers	Keeping Tadpoles (Alive!)	Science Dictionary	Materials: their Properties and Uses	My Holiday Diary	Transport Timelines	Arts and Crafts Around the World
Living things	✔	✔	✔					✔	✔					✔	✔	✔	✔					✔	✔						✔	✔	✔	✔					✔	✔	✔			
Local study				✔								✔									✔												✔	✔						✔	✔	
Machines																			✔	✔				✔								✔	✔	✔				✔	✔			
Materials										✔								✔							✔													✔	✔			✔
Movement								✔	✔																						✔	✔		✔							✔	
Myself		✔							✔								✔			✔		✔							✔											✔		
Number									✔																																	
Ourselves		✔				✔			✔								✔			✔		✔							✔		✔	✔								✔	✔	
Pattern and shape				✔										✔											✔																	✔
People who help us				✔	✔																✔						✔		✔													
Plants	✔	✔														✔							✔						✔													
School																					✔												✔	✔	✔							
Seasons	✔																																							✔		
Senses			✔						✔								✔																							✔		
Size											✔						✔																		✔							
Time													✔													✔							✔	✔	✔						✔	
Tools				✔															✔		✔				✔													✔				✔
Toys & games	✔				✔							✔									✔				✔	✔	✔						✔	✔								
Transport																													✔					✔							✔	✔
Vehicles																													✔					✔							✔	✔
Water										✔												✔	✔											✔						✔		
Weather	✔																						✔																			
Wheels																																									✔	

Using *Discovery World* to help teach subjects

The content of the *Discovery World* pupil books was based on extensive correlation to the National Curricula. We wanted to give you the right content coverage at the right level. Each pupil book has a main subject focus. You may like to photocopy this section and give a copy to each subject co-ordinator. We hope you will see that every title can be used to discuss many areas and develop many skills within each subject area. Literacy lessons which focus on key words are also a highly useful way to focus on the specialist vocabulary of each subject (words like 'melt', 'gravity', 'magnetic', etc.).

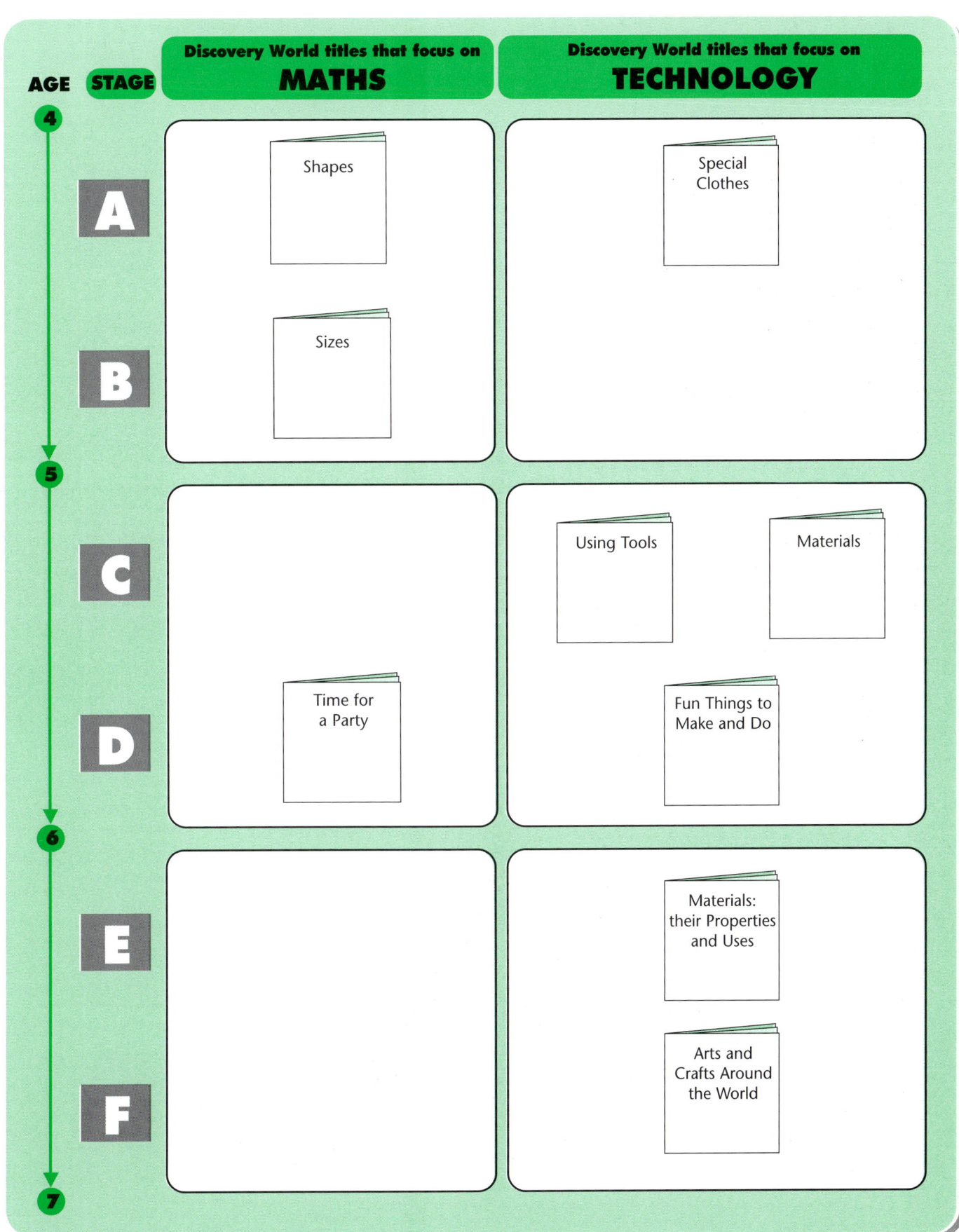

AGE	STAGE	Discovery World titles that focus on **MATHS**	Discovery World titles that focus on **TECHNOLOGY**
4	A	Shapes	Special Clothes
5	B	Sizes	
5	C		Using Tools Materials
6	D	Time for a Party	Fun Things to Make and Do
6	E		Materials: their Properties and Uses
7	F		Arts and Crafts Around the World

AGE | STAGE

Discovery World titles that focus on
SCIENCE

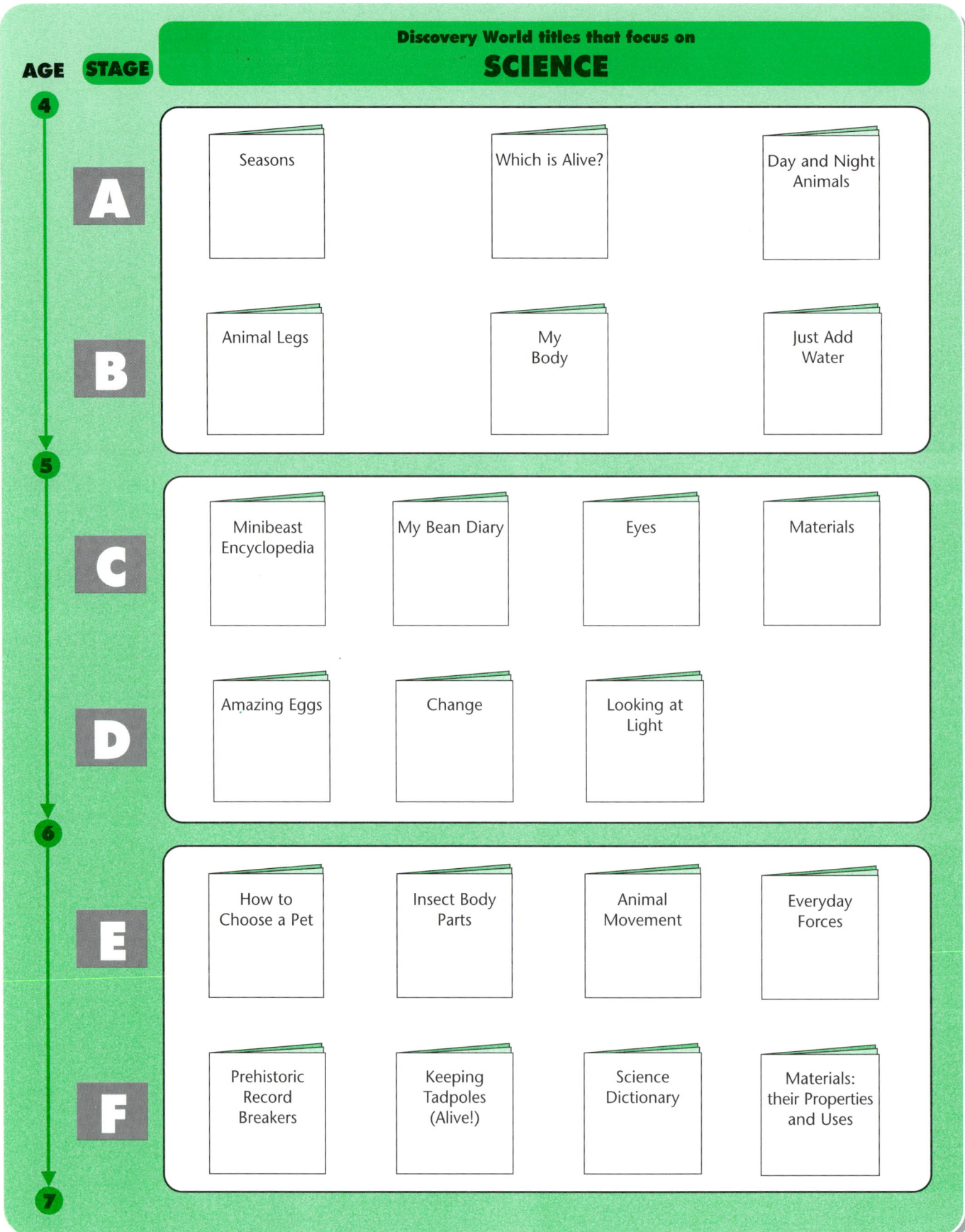

A

Seasons

Which is Alive?

Day and Night Animals

B

Animal Legs

My Body

Just Add Water

C

Minibeast Encyclopedia

My Bean Diary

Eyes

Materials

D

Amazing Eggs

Change

Looking at Light

E

How to Choose a Pet

Insect Body Parts

Animal Movement

Everyday Forces

F

Prehistoric Record Breakers

Keeping Tadpoles (Alive!)

Science Dictionary

Materials: their Properties and Uses

AGE: 4 5 6 7

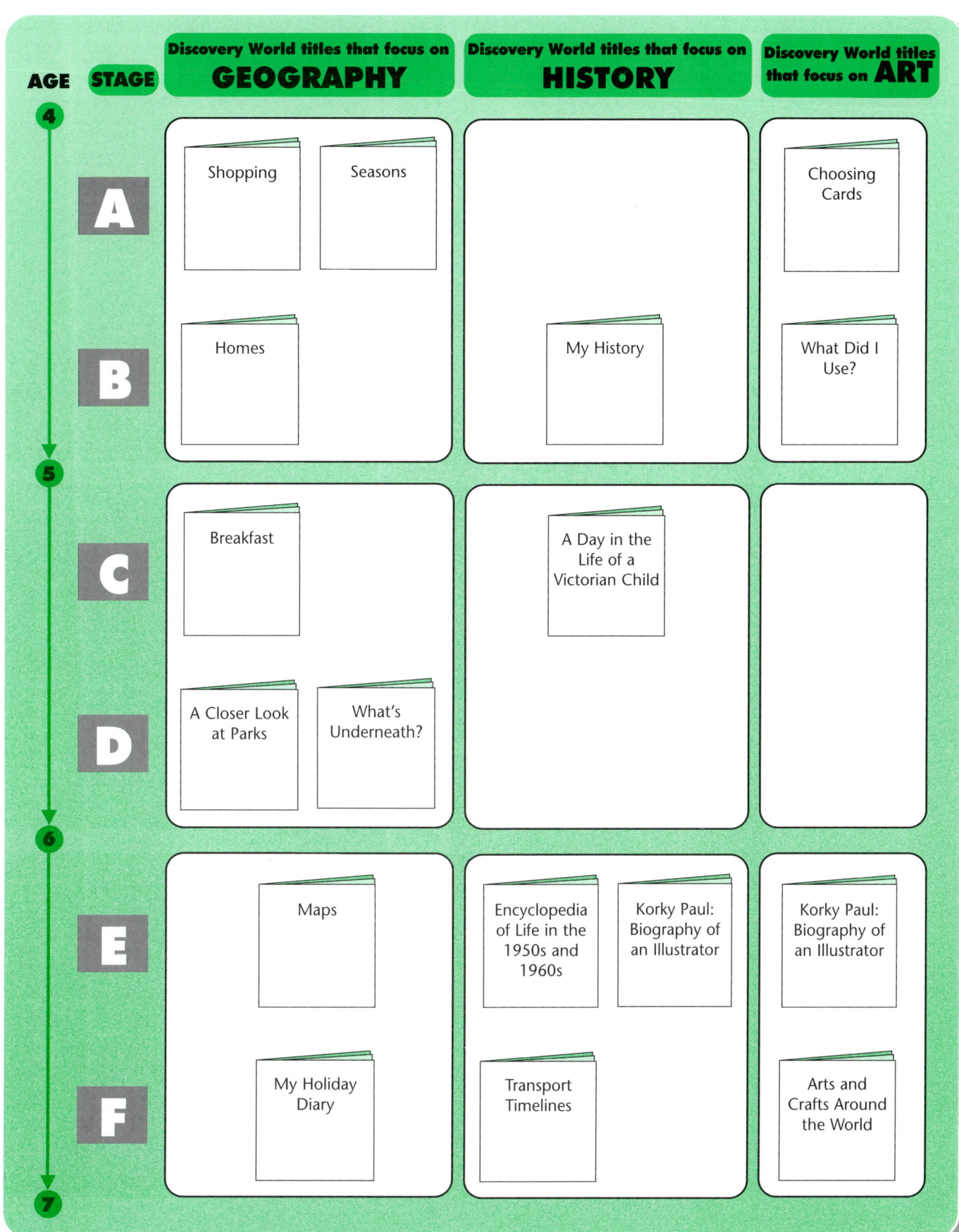

AGE	STAGE	Discovery World titles that focus on GEOGRAPHY	Discovery World titles that focus on HISTORY	Discovery World titles that focus on ART
4	A	Shopping · Seasons		Choosing Cards
5	B	Homes	My History	What Did I Use?
	C	Breakfast	A Day in the Life of a Victorian Child	
6	D	A Closer Look at Parks · What's Underneath?		
	E	Maps	Encyclopedia of Life in the 1950s and 1960s · Korky Paul: Biography of an Illustrator	Korky Paul: Biography of an Illustrator
7	F	My Holiday Diary	Transport Timelines	Arts and Crafts Around the World

© Reed Educational and Professional Publishing Ltd, 1997

Assessment and record-keeping

The following photocopiable sheets have been provided as aids to track non-fiction skill development. You can decide the level of individual monitoring you want to give to each child.

Assessment sheet	Purpose
1	To record lessons covered, and whether each child has shown evidence of mastering the learning objective, or whether the lesson needs revisiting.
2	A more detailed record than sheet 1. Records the skills covered, and whether each child has shown evidence of mastering the skill/learning objective, or whether the lesson needs revisiting.
3	Similar to sheet 2, but designed for the individual child with space for comments/diagnostic planning.
4	To record which book has been read by which child. This may be particularly useful to track which non-fiction books the children have taken home to share with a parent/carer.

This page can be used to record which lessons each child has been taught.

Needs revisiting – Fill in half the box

Mastered – Fill in complete box

NAMES

LITERACY LESSON BOOK 1

Lesson	Teaching Objective
1	We get information from photographs
2	We get information from print
3	We get information from books
4	The difference between fiction and non-fiction
5	How a non-fiction page works (1)
6	Understanding front covers
7	Non-fiction books are about different subjects
8	How a non-fiction page works (2)
9	How to 'dip in' or 'flick through'
10	How to ask questions
11	Identifying what you want to find out
12	Looking for key words
13	Introducing the contents page
14	Understanding simple charts
15	Understanding simple diagrams

LITERACY LESSON BOOK 2

Lesson	Teaching Objective
16	Assessment
17	Choosing the best book
18	Non-fiction books have different purposes
19	Using the contents page
20	Using the index
21	Reading and making diagrams (1)
22	Reading and making diagrams (2)
23	Reading and using a dictionary
24	Reading and using an encyclopedia
25	Reading and using a diary
26	Using the glossary
27	Using a library
28	Revision mini-topic (1)
29	Revision mini-topic (2)
30	Revision mini-topic (3)

This page can be used to record skills covered.

<u>Needs revisiting</u> – Fill in half the box

<u>Mastered</u> – Fill in complete box

NAMES

TEACHING OBJECTIVE/SKILL

READING AND USING THE DIFFERENT PARTS OF A NON-FICTION BOOK

- Understanding front covers
- Understanding what a 'double page spread' is
- Understanding page numbers
- Understanding page layout
- Understanding labels and captions
- Understanding page headings/sub-headings
- Using the contents page
- Using the index
- Using back cover blurbs
- Using a glossary

ESSENTIAL NON-FICTION READING SKILLS

- We get information from photographs (and pictures)
- We get information from print
- We get information from books
- The difference between fiction and non-fiction
- How a non-fiction page (book) works
- Non-fiction books are about different subjects
- How to 'dip in' or 'flick through' (early skimming & scanning)
- How to ask questions
- Identifying what you want to find out
- Looking for key words
- Choosing the best book
- Non-fiction books have different purposes
- Using a library
- Distinguishing between fact and opinion
- Understanding alphabetical order
- Early note-taking

READING AND CREATING CHARTS AND DIAGRAMS

- Understanding simple charts
- Understanding simple diagrams
- Making simple charts
- Making simple diagrams

READING AND WRITING DIFFERENT TYPES OF TEXT

- Simple environmental print
- Signs and notices
- Lists of instructions
- Atlas
- 'How to ...', instructional text
- Procedural text/recipes
- Biography
- Autobiography
- Interview
- Dictionary
- Encyclopedia
- Diary
- Report on a visit/trip

This page can be used to record skills covered.

NAME:

Teaching objective/skill	Introduced	Revisited	Mastered	Comments
READING AND USING THE DIFFERENT PARTS OF A NON-FICTION BOOK				
Understanding front covers				
Understanding what a 'double page spread' is				
Understanding page numbers				
Understanding page layout				
Understanding labels and captions				
Understanding page headings/sub-headings				
Using the contents page				
Using the index				
Using back cover blurbs				
Using a glossary				
ESSENTIAL NON-FICTION READING SKILLS				
We get information from photographs (and pictures)				
We get information from print				
We get information from books				
The difference between fiction and non-fiction				
How a non-fiction page (book) works				
Non-fiction books are about different subjects				
How to 'dip in' or 'flick through' (early skimming & scanning)				
How to ask questions				
Identifying what you want to find out				
Looking for key words				
Choosing the best book				
Non-fiction books have different purposes				
Using a library				
Distinguishing between fact and opinion				
Understanding alphabetical order				
Early note-taking				
READING AND CREATING CHARTS AND DIAGRAMS				
Understanding simple charts				
Understanding simple diagrams				
Making simple charts				
Making simple diagrams				
READING AND WRITING DIFFERENT TYPES OF TEXT				
Simple environmental print				
Signs and notices				
Lists of instructions				
Atlas				
'How to …', instructional text				
Procedural text/recipes				
Biography				
Autobiography				
Interview				
Dictionary				
Encyclopedia				
Diary				
Report on a visit/trip				

NAMES

This page can be used to record which books each child has read.

STAGE A

- Seasons
- Which is Alive?
- Day & Night Animals
- Special Clothes
- Shapes
- Shopping
- Choosing Cards

STAGE B

- Animal Legs
- My Body
- Just Add Water
- Sizes
- Homes
- My History
- What Did I Use?

STAGE C

- Minibeast Encyclopedia
- My Bean Diary
- Eyes
- Materials
- Using Tools
- Breakfast
- A Day in the Life of a Victorian Child

STAGE D

- Amazing Eggs
- Change
- Looking at Light
- Fun Things to Make and Do
- Time for a Party
- A Closer Look at Parks
- What's Underneath?

STAGE E

- How to Choose a Pet
- Insect Body Parts
- Animal Movement
- Everyday Forces
- Maps
- Encyclopedia of Life in the 1950s and 1960s
- Korky Paul: Biography of an Illustrator

STAGE F

- Prehistoric Record Breakers
- Keeping Tadpoles (Alive!)
- Science Dictionary
- Materials: their Properties & Uses
- My Holiday Diary
- Transport Timelines
- Arts and Crafts around the World

Photocopy masters

The following photocopy masters are designed to support the *Discovery World* *Literacy Lesson Books*. They are cleared for copying by the purchasing school only. Each photocopy master has information for the teacher explaining:

● which learning objective it reinforces,

● which literacy lesson it is connected to,

● which *Discovery World* book the child will need access to.

Many of these photocopy masters can be used for assessing the children's mastery of a learning objective.

This chart indicates which teaching objectives are supported by photocopy masters:

TEACHING OBJECTIVE	USE PHOTOCOPY MASTER
READING AND USING THE DIFFERENT PARTS OF A NON-FICTION BOOK	
Understanding front covers	8, 9, 24, 25
Understanding labels and captions	11
Using the contents page	18, 19, 30, 31
Using the index	32
Using a glossary	41
ESSENTIAL NON-FICTION READING SKILLS	
We get information from photographs (and pictures)	1
We get information from print	2, 3
We get information from books	4
The difference between fiction and non-fiction	6
How a non-fiction page (book) works	7, 12, 42
Non-fiction books are about different subjects	5, 10
How to 'dip in' or 'flick through' (early skimming & scanning)	13, 24
How to ask questions	14
Identifying what you want to find out	15, 43
Looking for key words	16, 17
Non-fiction books have different purposes	26, 27, 28, 29
Using a library	42
Distinguishing between fact and opinion	39
Understanding alphabetical order	36
Early note-taking	43
READING AND CREATING CHARTS AND DIAGRAMS	
Understanding simple charts	20, 21
Understanding simple diagrams	22, 23
Reading and making diagrams	22, 23, 33, 34, 35
READING AND WRITING DIFFERENT TYPES OF TEXT	
Procedural text/recipes	28, 29
Biography	27
Dictionary	36, 37
Encyclopedia	38, 39
Diary	40

Name

Is it in the picture? Write ✔ or ✗.

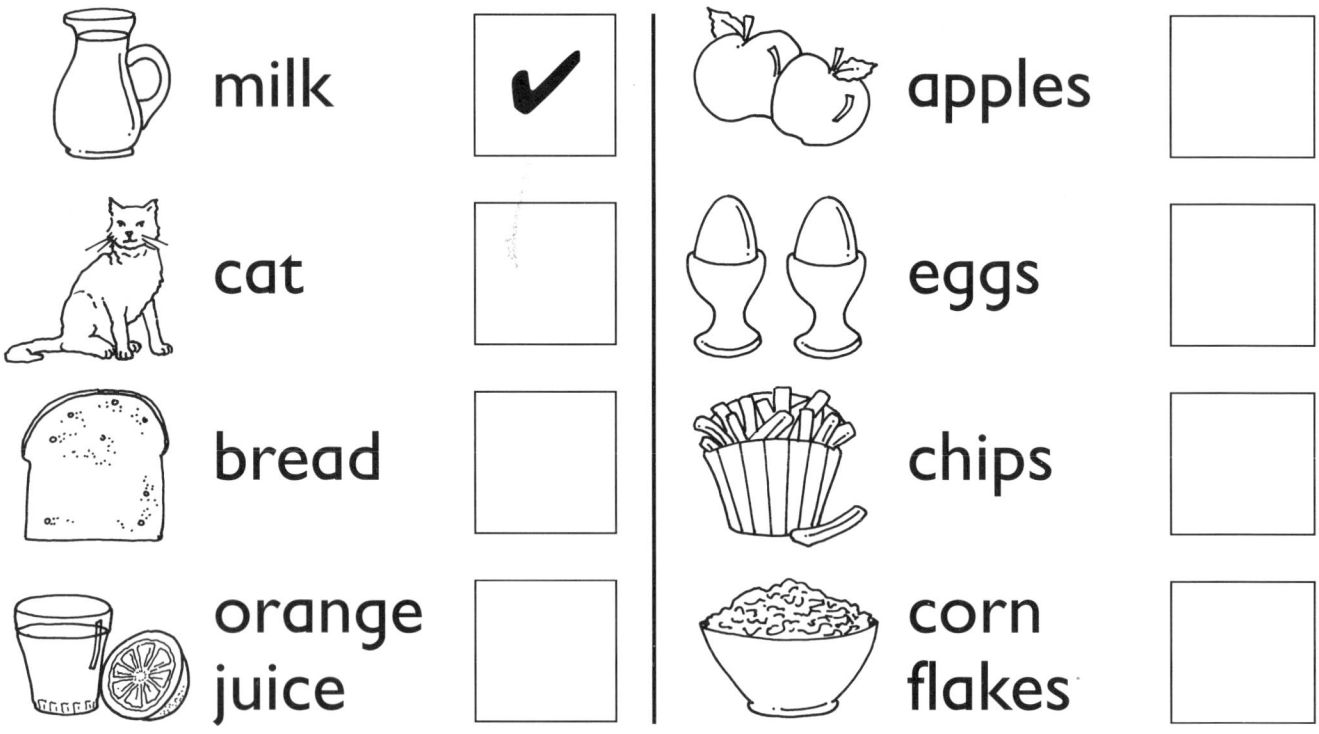

milk ✔   apples ☐

cat ☐   eggs ☐

bread ☐   chips ☐

orange juice ☐   corn flakes ☐

PCM 1

Teaching objective: **We get information from photographs (and pictures)**
Links with Literacy Lesson: **1**
Links with Discovery World: **Stage C** *Breakfast* **(page 2)**

© Reed Educational and Professional Publishing Ltd, 1997

Cut out the pictures. Match the pairs. Stick them on another piece of paper.

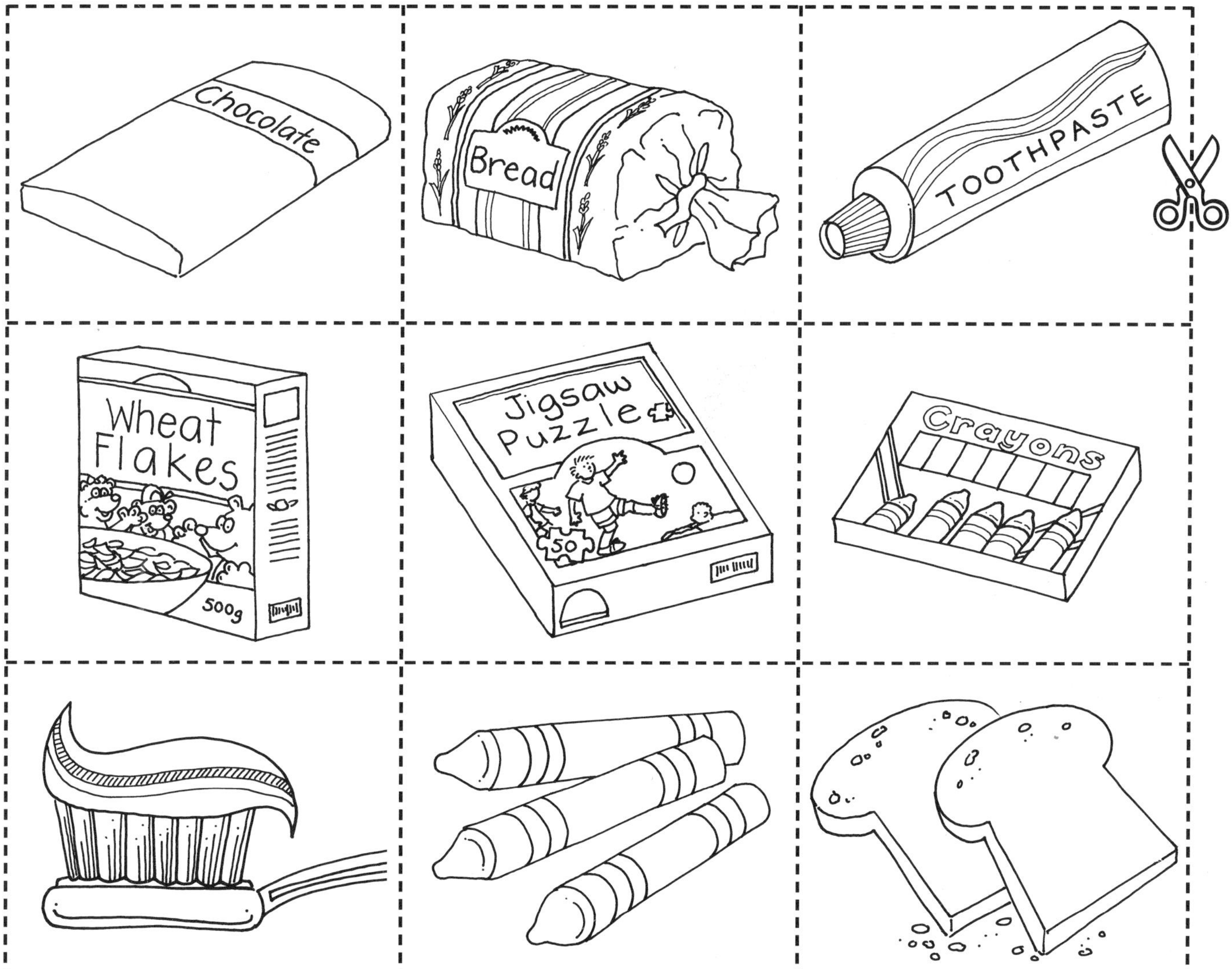

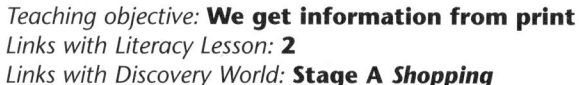

Teaching objective: **We get information from print**
Links with Literacy Lesson: **2**
Links with Discovery World: **Stage A** *Shopping*

© Reed Educational and Professional Publishing Ltd, 1997

Find the missing signs in picture 2.*

Teaching objective: **We get information from print**
Links with Literacy Lesson: **2**
Mark them with a cross or write them in according to ability.

© Reed Educational and Professional Publishing Ltd, 1997

Name

DISCOVERY
WORLD

Colour the circles yellow.

○ Colour the rectangles green.

▭

Colour the triangle red.

△ Colour the square blue.

▢

Teaching objective: **We get information from books**
Links with Literacy Lesson: **3**
Links with Discovery World: **Stage A** *Shapes*

Cut out the pictures. Match the pairs.
Stick them on another piece of paper.

Teaching objective: **Non-fiction books are about different subjects**
Links with Literacy Lesson: **4 and 7**

Name

Colour in the books which give us facts.

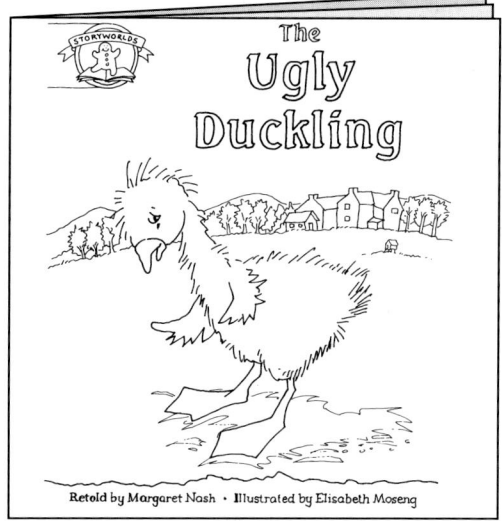

Teaching objective: **The difference between fiction and non-fiction**
Links with Literacy Lesson: **4**

Cut out the pictures.
Stick them on another piece of paper to make a non-fiction spread.*

What can you see in spring? ② ③

Teaching objective: **How a non-fiction page works**
Links with Literacy Lesson: **5**
Links with Discovery World: **Stage A *Seasons* (pages 2 and 3)**
*Children can use Literacy Lesson 5 pupil page as a model to follow
© Reed Educational and Professional Publishing Ltd, 1997

Name

Colour the title red.
Colour the author blue.
Colour the logo green.
Colour in the pictures.

DISCOVERY WORLD

Sizes

Jillian Powell

Teaching objective: **Understanding front covers**
Links with Literacy Lesson: **6**

PCM
8

Name

Design a front cover.
Think of what kind of book it will be.

Title

Logo

Author

Publisher

Teaching objective: **Understanding front covers**
Links with Literacy Lesson: **6**

© *Reed Educational and Professional Publishing Ltd, 1997*

Name

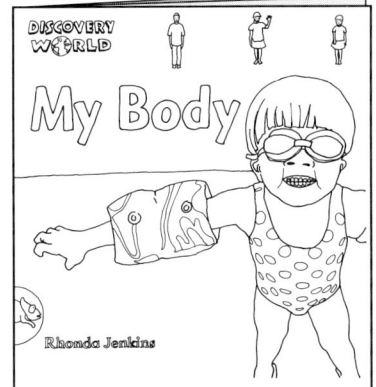

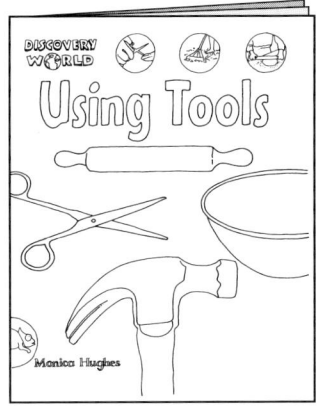

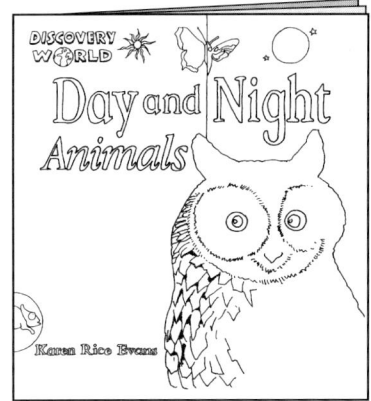

1 Which book is about animals?_____

2 Which book is about bodies?_____

3 Which book is about beans?_____

4 Which book is about shopping?_____

5 Which book is about tools?_____

6 Which book is about clothes?_____

7 Find another book about clothes.
Write the title: _____

Teaching objective: **Non-fiction books are about different subjects**
Links with Literacy Lesson: **4 and 7**

PCM
10

Name

Cut out the labels and stick by the picture.

| hat | boots | jacket | trousers | lollipop |

This is what a lollipop lady wears. ✂

Teaching objective: **Understanding labels and captions**
Links with Literacy Lesson: **8**
Links with Discovery World: **Stage A *Special Clothes***

© Reed Educational and Professional Publishing Ltd, 1997

Name

Non-fiction Book Report

The book I chose was

The author was

	Yes 😊	No 🙁
Contents page		
Page numbers		
Headings		
Photographs		
Illustrations (artwork)		
Charts or diagrams		
Labels		
Index		
Back cover blurb		

Teaching objective: **How a non-fiction page (book) works**
Links with Literacy Lesson: **8**

PCM
12

Flick through *Minibeast Encyclopedia* to find a fascinating fact.

Minibeast	Did you know...
	Ants leave a smell for other ants to follow.

Teaching objective: **How to 'flick through'**
Links with Literacy Lesson: **9**
Links with Discovery World: **Stage C Minibeast Encyclopedia**

My Interview

I would like to interview _____ .

These are the questions I want to ask:

1 Who _____ ?

2 What _____ ?

3 Why _____ ?

4 How _____ ?

5 When _____ ?

6 Where _____ ?

7 Which _____ ?

Teaching objective: **How to ask questions**
Links with Literacy Lesson: **10**

PCM
14

Name

We want to find out about

What we know K	What we want to find out W

Write down the answers to two of your questions.

1 _____ .

2 _____ .

Teaching objective: **Identifying what you want to find out**
Links with Literacy Lesson: **11**

Flick through a copy of *Fun Things to Make and Do.* Write and draw the things you need.

To make...	You need...

Teaching objective: **Looking for key words**
Links with Literacy Lesson: **12**
Links with Discovery World: **Stage D** *Fun Things to Make and Do*

Name

Look at a copy of *Minibeast Encyclopedia*.
You want to find out about *honeybees*.
Circle all the key words you can find below:

In the summer we can see many different kinds of insects. Butterflies land on flowers to feed on nectar. We hear the chirping sound of grasshoppers. Honeybees buzz around looking for flowers. They also feed on nectar and pollen. Honeybees can sting if attacked. After it stings the honeybee dies. Wasps can also sting, but they do not die afterwards.

PCM
17

Teaching objective: **Looking for key words**
Links with Literacy Lesson: **12**

© *Reed Educational and Professional Publishing Ltd, 1997*

Look at the contents page of *Looking at Light.*
On which page would you find out about:

- Electric light?   Page [ ]

- How the Sun makes light?   Page [ ]

- How matches makes firelight?   Page [ ]

- Different kinds of light?   Page [ ]

- Light at night from the Moon?   Page [ ]

- How batteries make light?   Page [ ]

Teaching objective: **Introducing the contents page**
Links with Literacy Lesson: **13**
Links with Discovery World: **Stage D** *Looking at Light*

© *Reed Educational and Professional Publishing Ltd, 1997*

Name

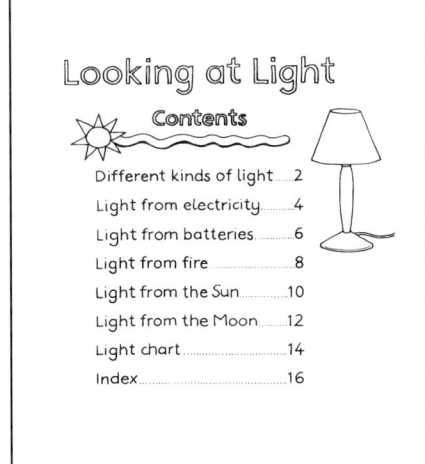

Looking at Light

Contents

Different kinds of light....2
Light from electricity.........4
Light from batteries.........6
Light from fire...............8
Light from the Sun.........10
Light from the Moon.......12
Light chart..................14
Index.........................16

Shapes

Contents

◯ Circle..............2
▢ Square............4
△ Triangle...........6
▭ Rectangle.........8
   Shapes...........10
   Index.............12

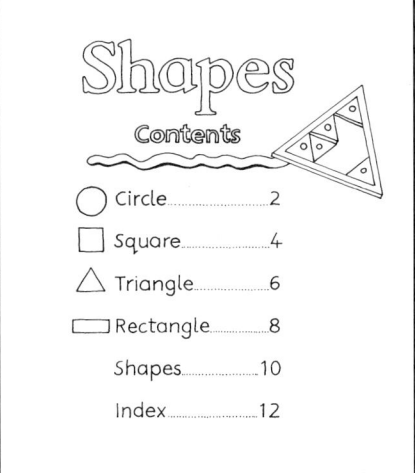

Breakfast

Contents

Breakfast.................2
Milk.......................4
Eggs......................6
Bread.....................8
Corn flakes..............10
Orange juice............12
Breakfast chart.........14
Index....................16

Look at these three contents pages.
Write down which book and which page you
need to look at to find out about these things.

	Book title	**Page**
Triangles		
Bread		
Moonlight		
Milk		
Electric light		

Teaching objective: **Introducing the contents page**
Links with Literacy Lesson: **13**

Look at a copy of *Animal Legs*. Draw the animals on the chart and label them. Think of some other animals and add them.

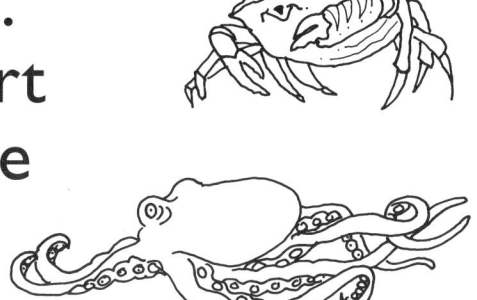

Number of Legs	Animals		
2	toucan	ostrich	human
4			
6			
8			
Lots			
None			

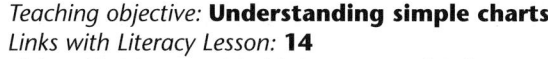

Teaching objective: **Understanding simple charts**
Links with Literacy Lesson: **14**
Links with Discovery World: **Stage B** *Animal Legs*

© *Reed Educational and Professional Publishing Ltd, 1997*

PCM
20

Cut up these animals and labels.
Use a copy of *Day and Night Animals* to help you make a chart like the one on page 10 and 11. Fill in the blank spaces.

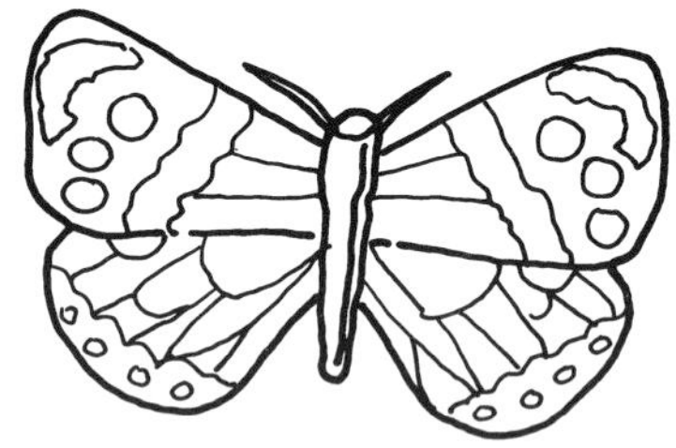

butterfly		blackbird
owl		moth
badger		ladybird

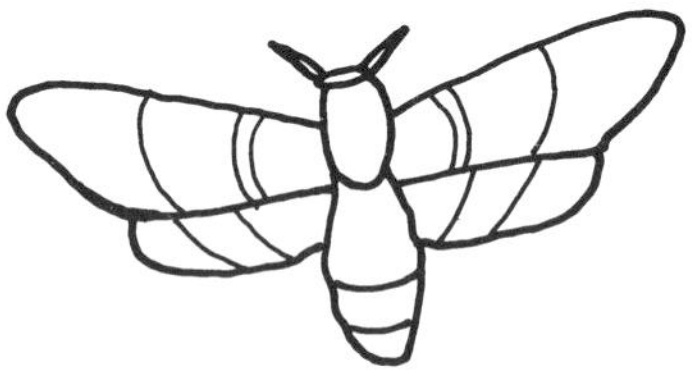

Teaching objective: **Understanding simple charts**
Links with Literacy Lesson: **14**
Links with Discovery World: **Stage A Day and Night Animals**

Look at a copy of *My Bean Diary*.
Label these diagrams.

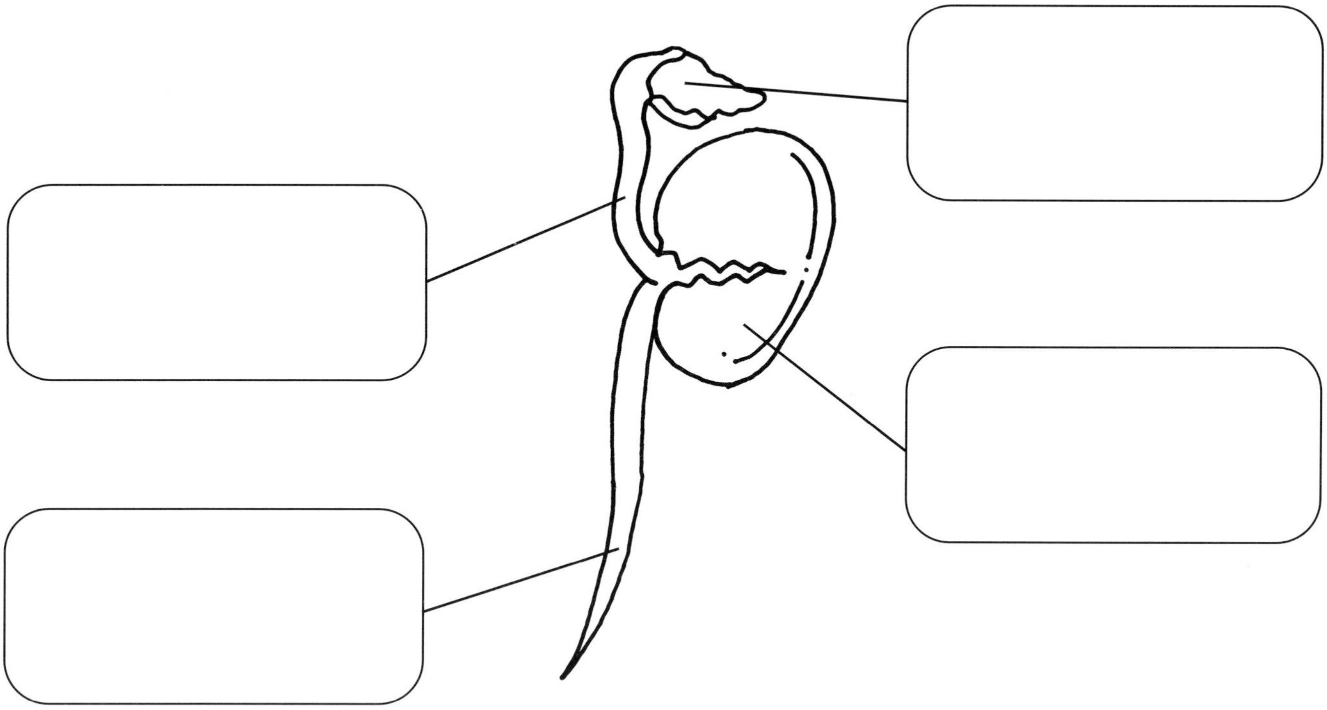

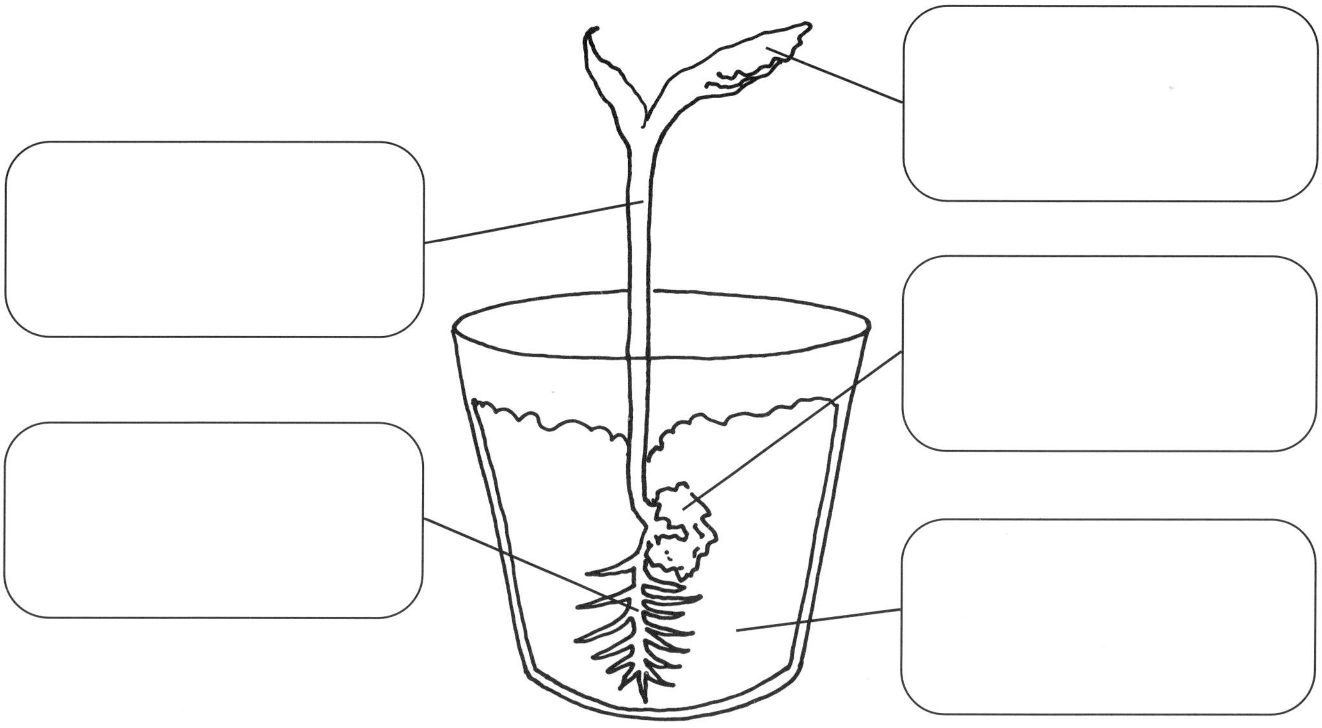

Teaching objective: **Labelling diagrams**
Links with Literacy Lesson: **15**
Links with Discovery World: **Stage C *My Bean Diary***

DISCOVERY WORLD

Label this diagram.
Use these words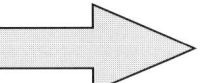

arm	head	chest
knee	hand	neck
leg	foot	fingers

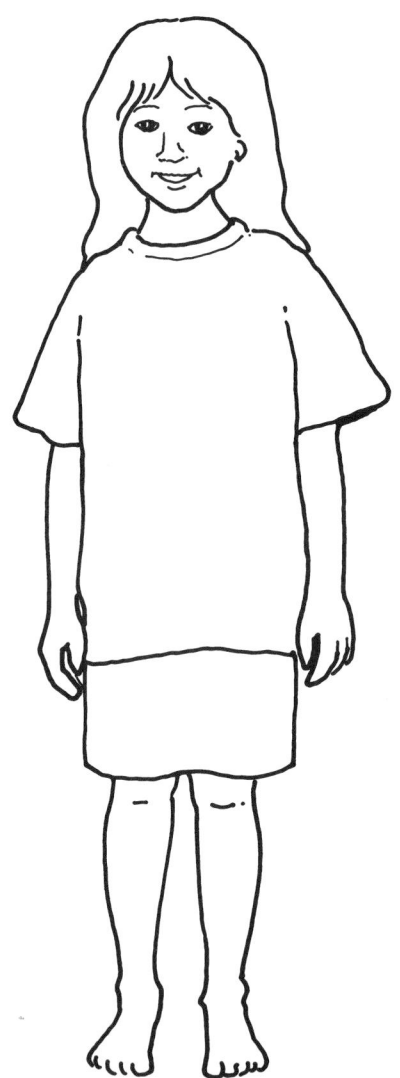

Can you add any more labels?

*Teaching objective: **Labelling diagrams***
*Links with Literacy Lesson: **15***
*Links with Discovery World: **Stage B My Body***

Name _____

Find some books about pets. Look at the books and fill in this chart. Put ✔ for YES. Put ✘ for NO.

Title and Author						

I think the best book to find out about _____

is _____ written by _____.

Teaching objective: **Understanding front covers**
How to 'flick through'
Links with Literacy Lesson: **16**

PCM
24

Cut up these front covers and back covers. Match them up and stick them on a new piece of paper.

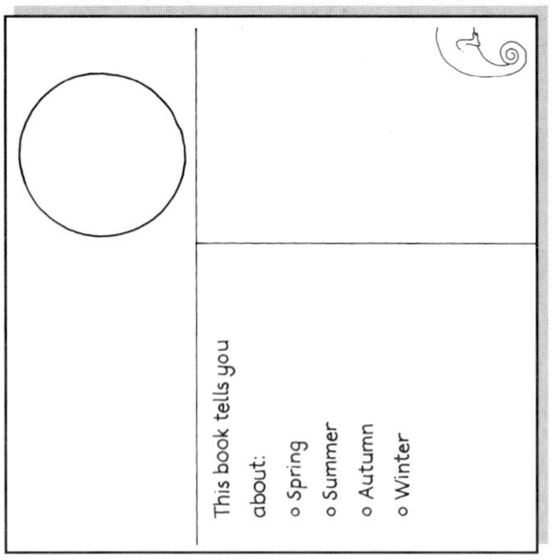

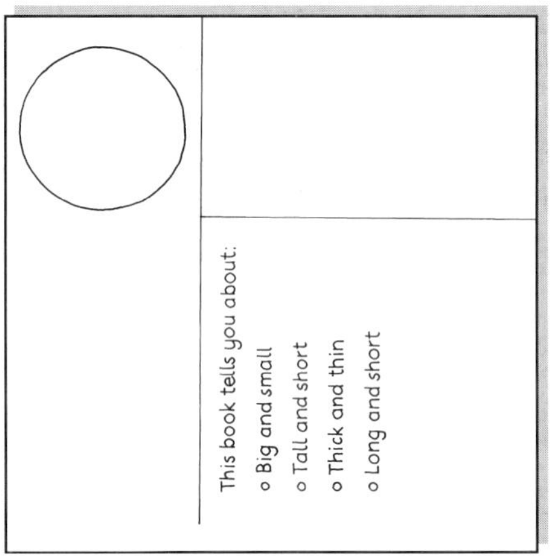

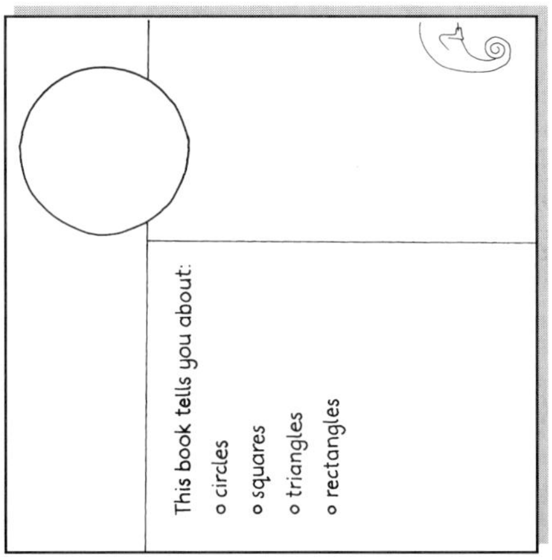

This book tells you
about:
- Spring
- Summer
- Autumn
- Winter

Shapes

DISCOVERY WORLD

Monica Hughes

This book tells you about:
- Big and small
- Tall and short
- Thick and thin
- Long and short

Sizes

DISCOVERY WORLD

Jillian Powell

This book tells you about:
- circles
- squares
- triangles
- rectangles

Seasons

DISCOVERY WORLD

Monica Hughes

*Teaching objective: **Understanding front covers***
*Links with Literacy Lesson: **17***

Atlas

Dictionary

Encyclopedia

Teaching objective: **Non-fiction books have different purposes**
Links with Literacy Lesson: **18**
Photocopy, cut out and mount on card. Use as 'flashcards', for display, signs, or to practise spelling.

© Reed Educational and Professional Publishing Ltd, 1997

Name

Planning a biography

What questions could you ask?

'Fact' questions	'Opinion' questions

'Fact' questions

When were you born?

Where do you live?

How many brothers and sisters do you have?

'Opinion' questions

Which is your favourite food?

Who is your best friend?

What books do you like?

Teaching objective: **Non-fiction books have different purposes (biography)**
Links with Literacy Lesson: **18**

Cut these out. Stick them on another piece of paper in the correct order with the numbers.

How to make a bed   1 2 3 4 5

Take the duvet and pillow off the bed.

Shake the pillow up and down.

Put the pillow back onto the bed.

Shake the duvet up and down.

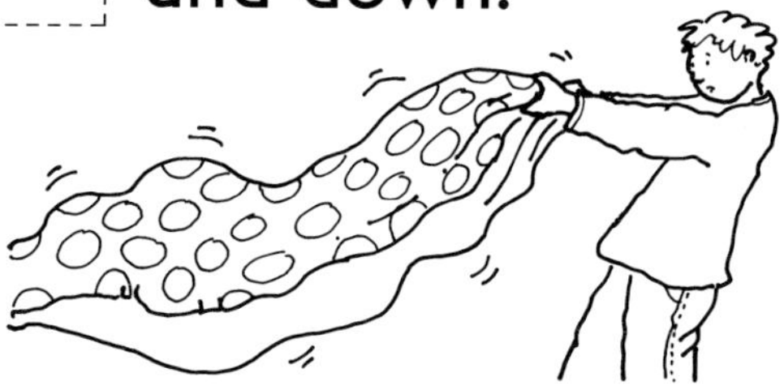

Put the duvet back onto the bed.

You will need:

One unmade bed.

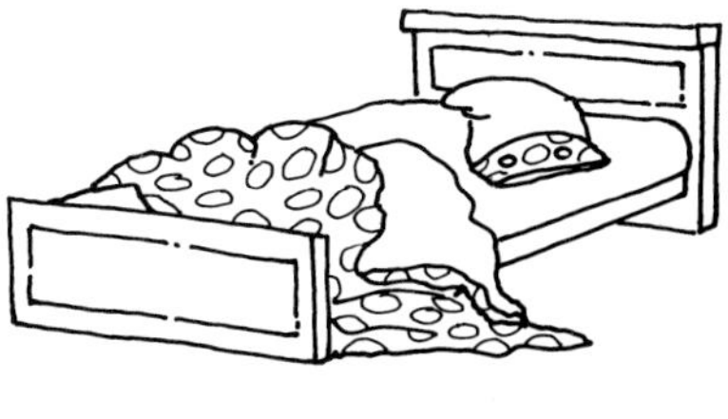

Teaching objective: **Non-fiction books have different purposes (writing a procedural text)**
Links with Literacy Lesson: **18**

Name

Read how a group of children made a milkshake. Underline in red the words which say what they **used**. Underline in blue the words which say what they **did**. Now write down all the things you need to make a milkshake here.

Today we made banana milkshake. We had a banana, $\frac{1}{2}$ pint (250 ml) of cold milk and some vanilla ice cream. Joe peeled the banana. I sliced it up. I used a knife. Sinead mashed it up in a bowl. She used a fork. Joe poured the milk into the bowl. I put 2 scoops of ice cream into the milk. Sinead mixed it all round. She used a whisk. It went all frothy. I poured it into 3 glasses. We all had a glass. Joe didn't like it so I had his.

How to make _____

You will need: _____

_____

_____

_____

Now write how to make a milkshake on another sheet of paper. Write numbers next to each thing you need to do, starting with number 1.

Teaching objective: **Non-fiction books have different purposes (writing a procedural text)**
Links with Literacy Lesson: **18**

Name

Question: ....

1 Book title	2 Contents page looks useful	3 Book *was* useful

Teaching objective: **Using the contents page**
Links with Literacy Lesson: **19**

Name

Look at each contents page. Write which page you would use to find out about these things.

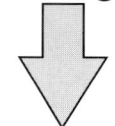

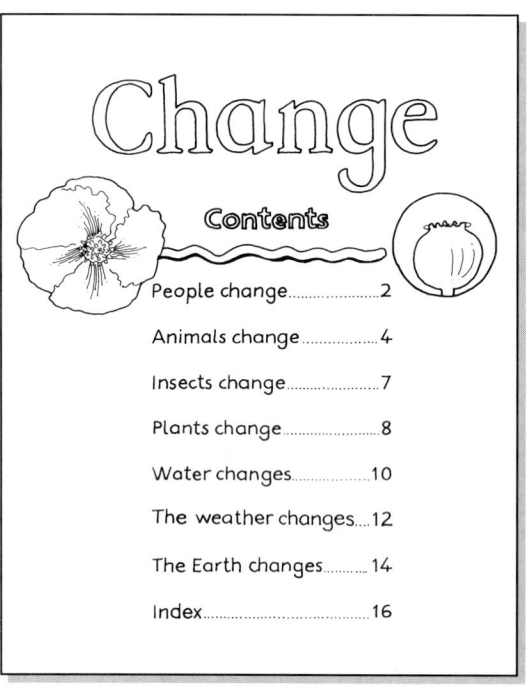

Change
Contents

a chameleon — page ....

a dragonfly — page ....

a poppy — page ....

a volcano — page ....

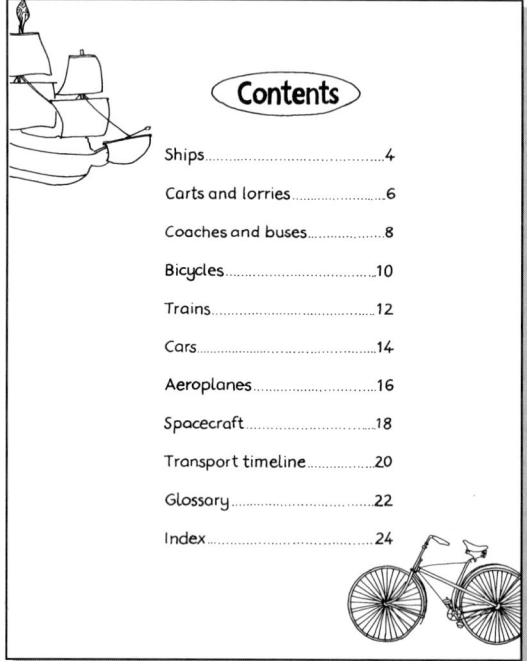

Contents

an oil tanker — page ....

a pennyfarthing — page ....

Viking longships — page ....

a stage coach — page ....

a diesel lorry — page ....

Teaching objective: **Using the contents page**
Links with Literacy Lesson: **19**
Links with Discovery World: **Stage D *Change* and Stage F *Transport Timelines***

© Reed Educational and Professional Publishing Ltd, 1997

This is the index page from a book called *A Day in the Life of a Victorian Child*. All the words are in the wrong order. Cut them out and stick them in the correct order on another piece of paper.

Index

a
b
c
d
e
f
g
h
i
j
k
l
m
n
o
p
q
r
s
t
u
v
w
x
y
z

16

candle **14, 15**

cooking range **4, 5**

jug **2, 3**

hoop **10, 11**

hot water bottle **14, 15**

penny **12, 13**

ladle **6, 7**

saucepan **4, 5**

scales **12, 13**

slate **8, 9**

slate pencil **8, 9**

spinning top **10, 11**

wash bowl **2, 3**

milk churn **6, 7**

Teaching objective: **Using the index**
Links with Literacy Lesson: **20**

DISCOVERY WORLD

Label this diagram. Answer the questions.

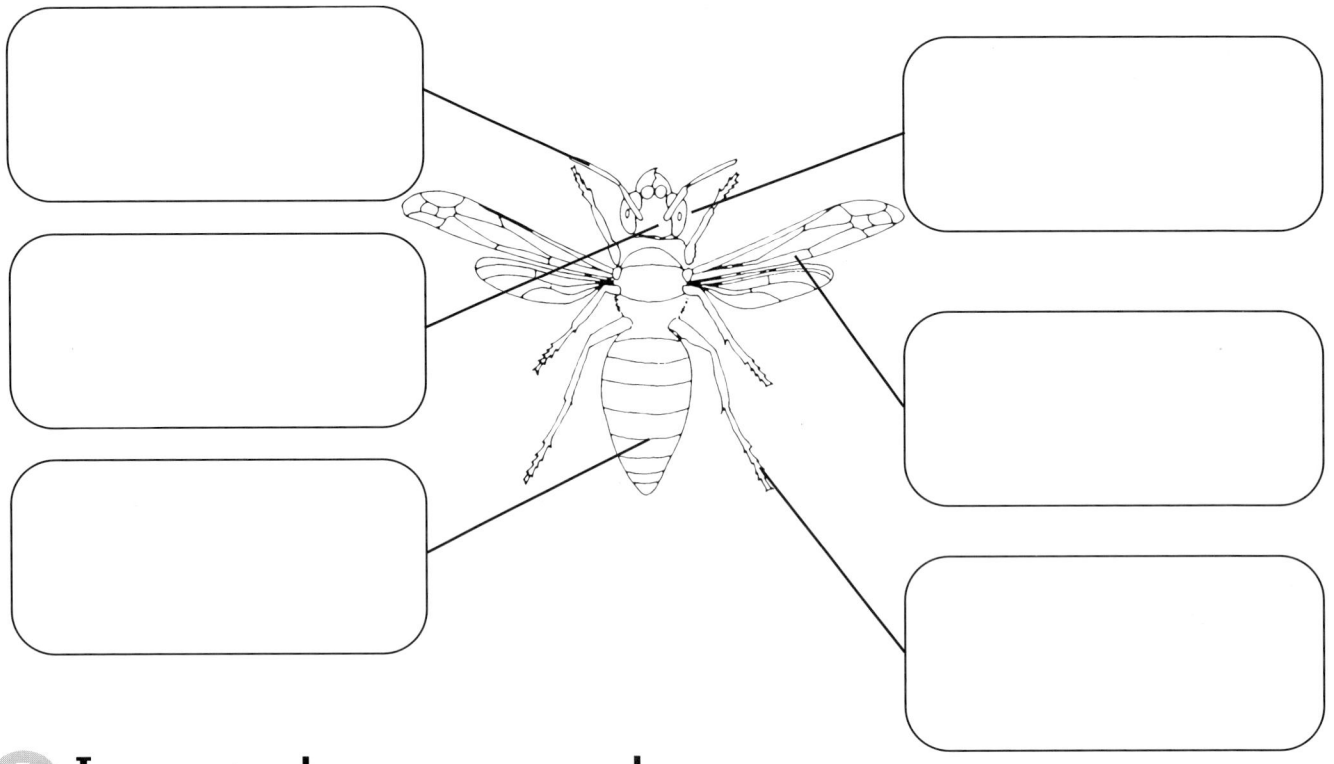

1 Insects have _____ legs.

2 Insects have a body with _____ main parts.

3 Insects have 'feelers' on their heads which are called _____ .

4 Is a crab an insect? _____

5 Is a fly an insect? _____

6 Is a spider an insect? _____

Teaching objective: **Reading and making diagrams**
Links with Literacy Lesson: **21**
Links with Discovery World: **Stage E** *Insect Body Parts*

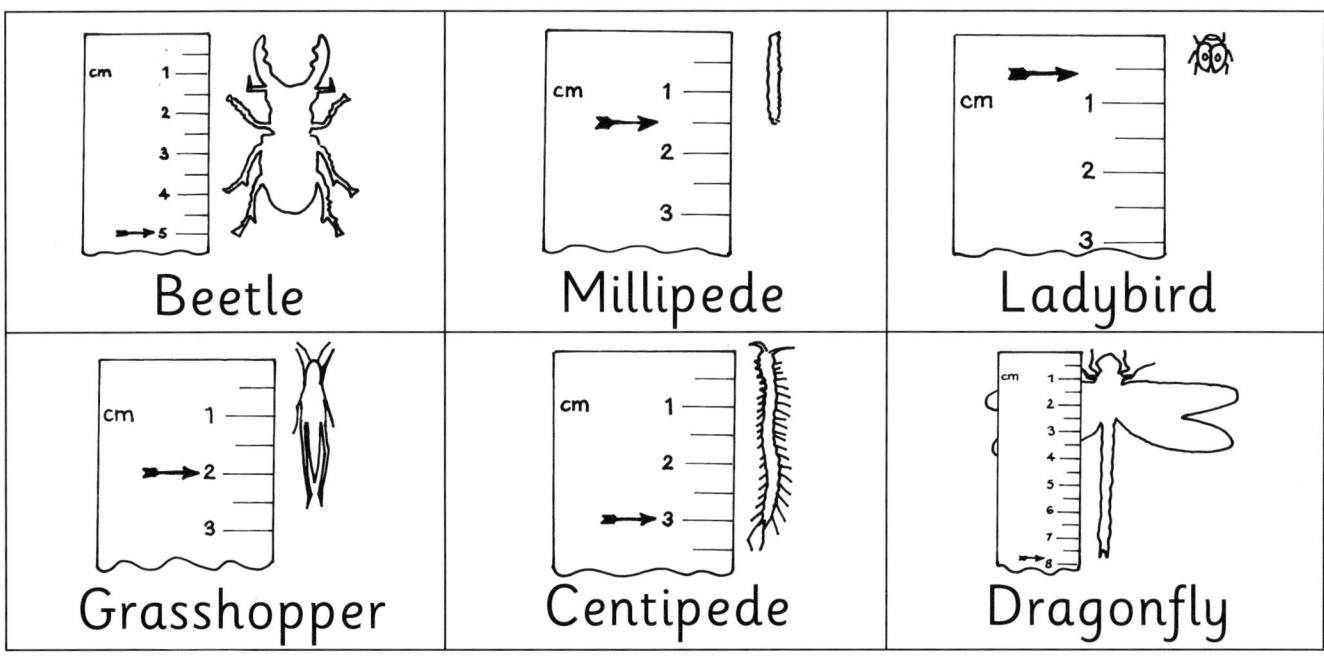

Beetle	Millipede	Ladybird
Grasshopper	Centipede	Dragonfly

Look at these minibeasts.
Which are less than 4 cms long?

_____

_____

_____

_____

Which are more than 4 cms long?

_____

_____

Teaching objective: **Reading and making diagrams**
Links with Literacy Lesson: **22**
Links with Discovery World: **Stage C Minibeast Encyclopedia**

© *Reed Educational and Professional Publishing Ltd, 1997*

PCM
34

Look at these diagrams and complete the sentences.

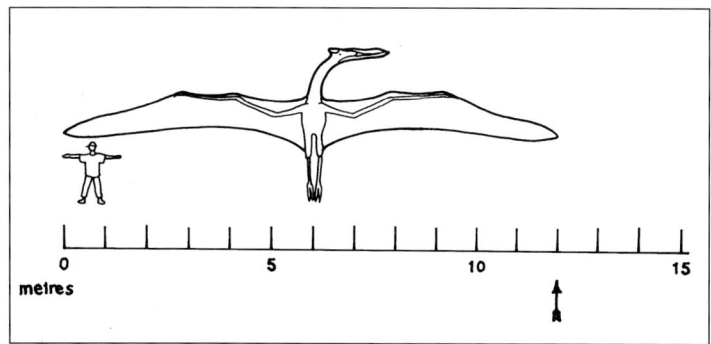

Quetzalcoatlus had a wingspan of _____ metres.

The biggest meat-eating dinosaur was

_____ .

It was_____
long.

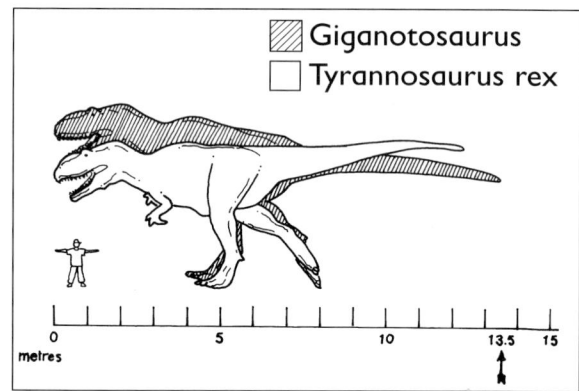

Giganotosaurus
Tyrannosaurus rex

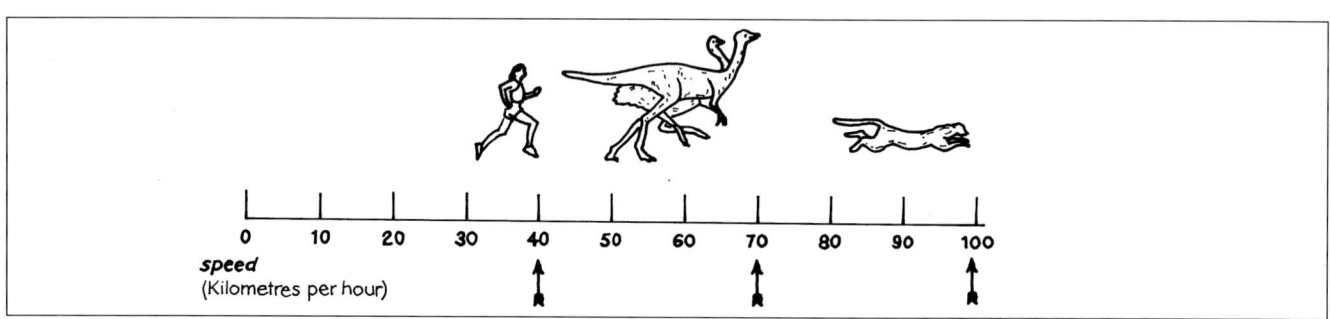

A cheetah can run at almost _____ kph.
Struthiomimus may have been able to _____
at _____ kilometres _____ hour.
Humans can run at about _____ .

Teaching objective: **Reading and making diagrams**
Links with Literacy Lesson: **22**
Links with Discovery World: **Stage F Prehistoric Record Breakers**

© Reed Educational and Professional Publishing Ltd, 1997

Cut out these boxes. Arrange in alphabetical order. Which letters are missing? Write a sentence about each animal.

Cc cat	Rr rabbit	Dd dog	Ii iguana
Aa ant	Ss sealion	Ff fox	Oo owl
Hh horse	Gg goat	Zz zebra	Tt tiger
Vv vulture	Mm monkey	Yy yak	Qq quail

Teaching objective: **Reading and using a dictionary (alphabetical order)**
Links with Literacy Lesson: **23**

PCM
36

Cut out these boxes. Make a dictionary page. Match each word to its meaning. Which guide letters will you write at the top of the page?

paper	to make something liquid when heated
plastic	a material that is shiny and strong and a good conductor of heat and electricity
melt	a long case of a plant that holds seeds
plant	a thin material made from wood pulp
pupil	an animal that is kept and cared for at home or at school
pod	a man-made material that is light and easy to bend
nest	a place where some animals lay their eggs
pet	a living thing that makes its own food from water, air and sunlight
metal	the black part of an eye that lets light through

Teaching objective: **Reading and using a dictionary**
Links with Literacy Lesson: **23**

Name

Use a copy of *Minibeast Encyclopedia* to complete this chart.

Name and illustration	Habitat (where it lives)	Diet (what it eats)	STAR FACT

Teaching objective: **Reading and using an encyclopedia**
Links with Literacy Lesson: **24**
Links with Discovery World: **Stage C Minibeast Encyclopedia**

Use *Encyclopedia of Life in the 1950s and 1960s* to write reasons why these people might have thought these things.

I think the 1950s were best because...

I think the 1960s were best because...

I think **now** is best because...

Teaching objective: **Reading and using an encyclopedia**
Distinguishing between fact and opinion
Links with Literacy Lesson: **24**
Links with Discovery World: **Stage E Encyclopedia of Life in the 1950s and 1960s**
© *Reed Educational and Professional Publishing Ltd, 1997*

Use *My Holiday Diary* to complete this diary for Jason. What do you think he did on Saturday?

Jason's Holiday Diary

On Sunday ....

....

On Monday ....

....

On Tuesday ....

....

On Wednesday ....

....

On Thursday ....

....

On Friday ....

....

On Saturday ....

....

Teaching objective: **Reading and using a diary**
Links with Literacy Lesson: **25**
Links with Discovery World: **Stage F** *My Holiday Diary*

PCM
40

This is the glossary page from a book called *Prehistoric Record Breakers.* Match the glossary words to the definitions. Stick them in the correct order on another piece of paper.

Glossary

	an animal that hunts and eats other animals
scale diagram	an animal that is hunted and eaten by other animals
fossil	
wingspan	the length of an animal from the tip of one wing to the tip of the other wing
prey	
predator	a drawing that shows the size of something
	the remains of a living thing that has turned to stone

Teaching objective: **Using the glossary**
Links with Literacy Lesson: **26**

Non-fiction Book Review

Title:

Author:

Organisation	YES 😊	NO 😟
Contents page		
Useful headings		
Glossary		
Index		
Useful back cover blurb		

I thought this book was .....

....

....

....

....

Style	YES 😊	NO 😟
Easy to read		
Interesting		
Clear		

Photographs	YES 😊	NO 😟
Interesting		
Useful		
Labels/captions		

Illustrations	YES 😊	NO 😟
Interesting		
Useful		
Labels/captions		

Teaching objective: **Using a library**
           How a non-fiction page (book) works
Links with Literacy Lesson: **27**

PCM
42

What do we know? K	What do we want to know? W	What we have learned L

Teaching objective: **Identifying what you want to find out**
Early note-taking
Links with Literacy Lesson: **30**